INSECTS *of the* PACIFIC NORTHWEST

Peter Haggard & Judy Haggard

All photos by Peter Haggard.

Half title page: Ceanothus silk moth (*Hyalophora euryalus*), male. Frontispiece: Sheep moth (*Hemileuca eglanterina*), male. This page: *Disonycha alternata,* mating pair with male on top. Page 6: Rathvon's scarab (*Lichnanthe rathvoni*). Page 8: Omnivorous looper (*Sabulodes aegrotata*) on vine maple leaves. Page 20: Common whitetail (*Libellula lydia*), immature male. Page 262: Shamrock spider (*Araneus trifolium*) in silken lair.

Published in 2006 by
Timber Press, Inc.
The Haseltine Building
133 S.W. Second Avenue, Suite 450
Portland, Oregon 97204-3527
timberpress.com

2 The Quadrant
135 Salusbury Road
London NW6 6RJ
timberpress.co.uk.

Designed by Susan Applegate
Printed in China

Fifth printing 2013

Library of Congress Cataloging-in-Publication Data

Haggard, Peter.
Insects of the Pacific Northwest/Peter Haggard and Judy Haggard.
p. cm.
Includes bibliographical references and index.
ISBN-13: 978-0-88192-689-7
1. Insects—Northwest, Pacific—Identification.
I. Haggard, Judy. II. Title.
QL475.N9H34 2006
595.7'09795—dc22 2005021973

A catalog record for this book is also available from the British Library.

Contents

Acknowledgments

We wish to extend gratitude and thanks to the following people:

W. J. Holland, M. Hatch, and E. O. Essig, whose books are the foundation on which this field guide is written. Without the information they provide, we would never have been able to start on this journey;

Richard Hurley, PH.D., Professor Emeritus, Humboldt State University, and Affiliate Professor and Associate Curator of Diptera, Montana State University, Bozeman, Montana, who provided Pete access to Humboldt State University's entomology laboratory and equipment and helped him use insect keys long after he graduated;

Norman Penny, PH.D., Senior Collections Manager, and the staff at The California Academy of Sciences, San Francisco, California, who were generous with their help. The Academy's insect collection has been an invaluable source for identifying insect species and confirming their ranges;

George Poinar, PH.D., for generously giving Pete the opportunity to use the insect collection at Oregon State University, Corvallis, Oregon;

The staff of the Plant Pest Diagnostics Laboratory, California Department of Food and Agriculture, Sacramento, California, for identifying the many pest insects Pete has sent them during the 33 years he has worked for the Humboldt County (California) Department of Agriculture;

Ken Hansen, whose help has been invaluable in identifying the butterflies and larger moths and for providing Pete with moths at various life stages that he would never have been able to obtain on his own;

Dick Penrose and Larry Bezark, long-horned beetle experts, who identified or confirmed the identification of the many long-horned beetle specimens;

Kathy Biggs, dragonfly expert, for identifying the dragonflies on photograph slides;

Steven Darington and Nezzie Wade, who gave much of their time to help edit the manuscript. Steven provided sound advice and computer technical assistance that made editing this field guide much easier;

Rosada Martin and Marsha Mello, for the illustration of the external structure of insects;

Neal Maillet, Executive Editor at Timber Press, for being so patient.

Introduction

Insects have intrigued Pete since his childhood days. Over the years, his interest in them grew, and, in 1990, he started photographing them. Since then, he has amassed a large collection of insect photographs along with an extensive database of information about their life histories, constructed from the many field notes he has taken. During this time he has also raised hundreds of insect species and thousands of individuals in order to confirm their identification and photograph their different life stages. In researching the literature on insects, he found that there are very few insect field guides that cover the various geographical regions within the United States despite the fact that there are hundreds of thousands of insect species in the United States (millions worldwide). Thus was born the idea for this field guide.

This book is intended as an introductory guide and natural history of insects of the Pacific Northwest, with its primary focus the identification of insects. It has been written with the non-scientist in mind. For those readers who are unfamiliar with entomology, that is, the study of insects, we have included sections on scientific nomenclature and classification, insect anatomy, and insect growth and development. Although very brief, these sections, along with the section describing the layout of this field guide and the glossary, should provide the reader with enough basic background information to deal with the individual insect accounts herein. If the reader wishes to obtain more general information regarding insects, there are many good textbooks available that discuss these and other topics in greater detail (see the bibliography).

Classification and Nomenclature

Scientists classify organisms, that is, they arrange them into groups, primarily on the basis of common structural characteristics (which may not reflect evolutionary relationships). This arrangement, called the Linnaean system of classification, consists of a hierarchy of categories called *taxa* (singular, *taxon*), with each higher taxon containing one to many taxa of the next subordinate rank. The principal categories in the classification of organisms are, in descending sequence, phylum, class, order, family, genus, and species. Intermediate categories (for example, superfamily and subspecies) are often used.

Each species has a scientific name, which is based on the two lowest taxa, genus (plural, *genera*) and species (plural, *species*). Scientific names are Latinized, and many are derived from Latin or Greek words. The classification and scientific name of a species may change as more information, particularly genetic (DNA), is learned about it and its relation to other species.

In addition to its scientific name, a species may have a common name. Common names are colloquial in nature, that is, no formal rules govern them. They are less precise than scientific names; several common names may apply to the same species, and the name(s) may vary from region to region. Many species do not have a common name. This is especially true for invertebrate animals, a group that is composed of a huge number of species of which many are seldom encountered or poorly known.

What Is an Insect?

Insects (class Insecta) are invertebrate animals that belong to the phylum Arthropoda, along with other invertebrates such as spiders, crustaceans, and millipedes. All arthropods have in common a supportive outer covering, called an *exoskeleton*, and a segmented body with jointed appendages.

Members of the class Insecta are distinguishable from other arthropods by having three distinct body regions—head, thorax, and abdomen—and no more than three pairs of legs. The typical external features of insect anatomy are shown in the illustration.

The head of an insect bears the antennae, eyes, and mouthparts. The pair of antennae are used for touching and smelling, and, in some insects, for hearing. Most insects have two types of eyes—multi-faceted compound eyes and light-sensitive simple eyes or *ocelli*. The mouth is surrounded by several structures collectively known as the mouthparts. The principal part usually consists of a pair of jaws, or *mandibles*, used for biting and chewing, but these structures are modified into an elongated proboscis or beak for lapping, or piercing and sucking in some insect groups.

The thorax, or middle section of an insect's body, bears the legs and wings (when present). Many insects, particularly the bugs and the beetles, have a *pronotum*, the upper surface of the part of the thorax that extends from the head to the base of the wings, and a *scutellum*, a distinct, triangular area located immediately posterior to the pronotum. The legs are each divided into three main segments: an elongate femur (thigh); a slender tibia (shin); and a terminal tarsus (foot), which may end in claws. Most adult winged insects have two pairs of wings. At least one pair is membranous and contains a network of thickened ridges, called *veins*, although the wings may be greatly modified.

The abdomen usually consists of eleven segments, although some insects have fewer. All insects have an abdomen that has pairs of breathing pores, or *spiracles*, along the sides (in addition to the one or two pairs of spiracles on the thorax). The abdomen may have a pair of sensory appendages at the tip, called *cerci*. And, in many species, the female has a structure at the end of her abdomen, called an *ovipositor*, which she uses to lay her eggs. Gills, present in the immature stages of some aquatic insects, are located on the abdomen.

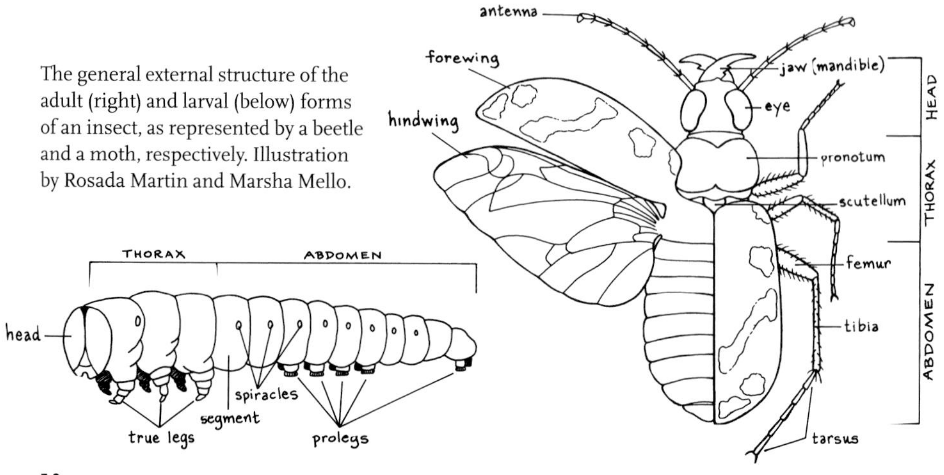

The general external structure of the adult (right) and larval (below) forms of an insect, as represented by a beetle and a moth, respectively. Illustration by Rosada Martin and Marsha Mello.

Insect Growth and Development

The young of many insect species hatch from eggs after they have been laid. As an immature insect grows, it must shed, or molt, its exoskeleton (as does a spider) because the exoskeleton is limited in its ability to stretch. Growth occurs through a series of stages, called *instars*; at the end of each instar, the exoskeleton is shed, and the insect enlarges. The number of molts varies with each species. The developing insect may not only change its size but also its body form and habits through a process of change known as *metamorphosis*. Although there is much variation in this process between the different insect taxa, these variations can be grouped into two general types, *complete metamorphosis* and *incomplete*, or *simple*, *metamorphosis*.

The majority of species undergo complete metamorphosis, that is, they pass through four stages of development—egg, larva, pupa, and adult. In their immature versus their adult stages, insects are usually quite different in form, have very different habits, and often live in different habitats. The larval stage is the active feeding stage in an insect's development; it is especially adapted to finding and consuming food. In addition to thoracic legs, the larvae of many species have abdominal legs; others are legless. The wings (if any) develop internally within the larva. Following the last instar molt, the insect transforms into the pupal stage. This is a "resting" stage in that the insect does not feed and usually appears on the surface to be inactive; however, considerable structural reorganization is taking place within as the larva transforms into the adult. The pupae of numerous species are enclosed in a cocoon or some other type of covering, and many pass the winter as a pupa. The final molt occurs at the end of the pupal stage, resulting in the emergence of a fully developed adult.

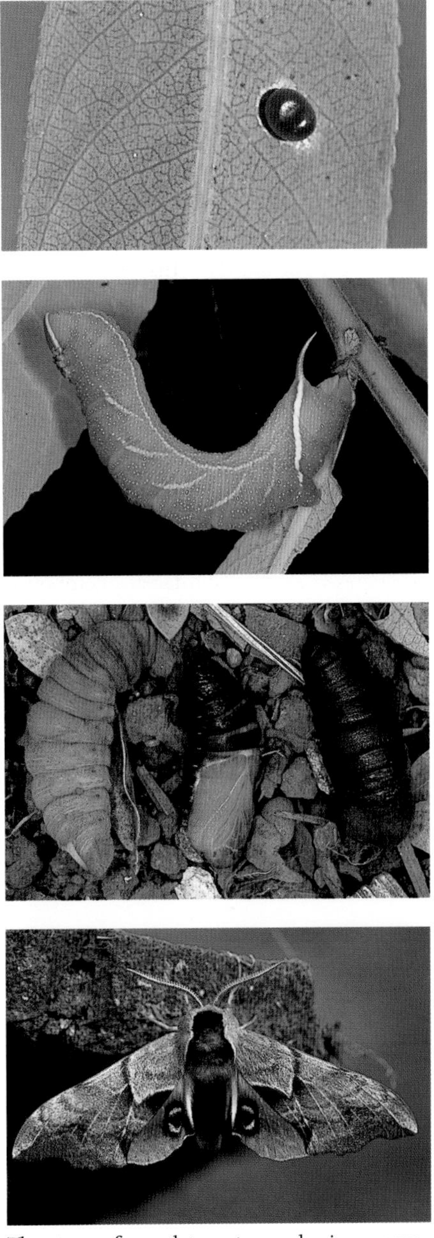

The stages of complete metamorphosis, as represented by *Smerinthus cerisyi* (eyed sphinx): egg, larva, pupa (early, middle, and final stage), and adult.

Other insects undergo incomplete, or simple, metamorphosis, in which there are three stages of development—egg, nymph, and adult. The nymphal stage is equivalent to the larval stage of those insects with complete metamorphosis in that it is the major feeding stage in development. And, like larvae, nymphs lack functional wings. Any wings that develop, however, do so externally; nymphs usually have wing pads which gradually enlarge and develop into wings as the nymph grows. Otherwise, in most insects that undergo incomplete metamorphosis, the nymph is usually similar to the adult, undergoing relatively slight to moderate change in body form and habits during development, and both live in the same habitat. In some orders, though, the nymphs do differ considerably from the adult stage: they are aquatic and breathe by means of gills, while the adults are winged and airborne. These nymphs are sometimes called *naiads*. For all insects with incomplete metamorphosis, there is no pupal stage preceding the last molt.

The stages of incomplete metamorphosis, as represented by the stink bug *Elasmostethus cruciata:* eggs and adult (female), first instar nymphs (the newly emerged nymphs are completely red), and last instar nymph.

Insect-induced Plant Galls

We have included many insect-induced plant galls in this field guide because, although they are rarely represented in other field guides, they are a common component of most plant communities (they are particularly abundant in oak woodlands). Galls are abnormal growths of plant cells that most commonly occur on leaves and stems but can occur almost anywhere on a plant. Plant galls come in a variety of sizes, colors, and shapes —from a simple swelling to odd and amazing shapes like stars, cones, and hollow spheres—each form unique to the species of insect.

The formation of a gall is started by the chemical and/or mechanical stimulus of an attacking insect or plant. Since this is a field guide on insects, the only galls included here are those induced by insects. The gall is created as a result of chemical manipulation of the plant by the insect and/or a reaction by the plant to the invading insect. In the case of chemical stimulation, a female laying her eggs and/or a larva developing within the plant produce chemicals that may be rich in nucleic acids that instruct the plant to produce a particular form of gall for that species. In some cases, the plant itself may contribute to the formation of a gall by trying to wall off the invasion, similar to an immune response by our cells in fighting off invading viruses or bacteria. In the process, the plant tissue expands near the invasion site, and in doing so, forms a gall that will house and provide nourishment for the growing larva(e). The gall normally causes little damage to the host plant. The larva(e) that lives inside the gall is protected from direct exposure to predators and parasites; however, there is a group of predators and parasites that have evolved to prey on larvae within galls by breaching the gall.

The insects that induce plant galls are usually wasps and flies, although there are some species in other orders as well as mites that also cause galls to form. Although the galls themselves are easy to identify (there are several books on plant galls that have good pictures), all the wasps and flies that induce the galls are tiny and difficult to identify and there are no identification keys geared to the amateur entomologist. The life cycle of some of the wasp species are complex: they produce two generations per year, each generation producing a differently shaped gall and one of the generations reproducing parthenogenetically. Some gall wasps induce the plant to produce honeydew, which may be a significant source of food for nectaring insects such as butterflies, bees, and wasps. The insects that are attracted to honeydew may, in turn, protect the gall from being attacked by predaceous and parasitic insects.

About This Guide

This field guide describes insects that occur in the Pacific Northwest, from southwestern British Columbia to northern California. Most of the species included are native to the Pacific Northwest region; the few that are not native are indicated as introduced in their accounts. Since insects do not recognize state or national boundaries, most of the insects in this book have ranges outside the Pacific Northwest.

There are at least 28,000 (and still counting) species of insects in the Pacific Northwest. We have included in this introductory guide 452 species in 10 orders (there are 30 orders worldwide, but not all occur in the region). And although this is basically a book about insects, we have also included 19 non-insect terrestrial invertebrate species for a sampling of interesting small creatures other than insects that the reader might likely encounter. The criteria Pete used to select the insects were that they had to be (1) common (a species likely to be seen by many people), (2) large enough to photograph well, and/or (3) distinct enough that they can be identified by a photograph. He also took into consideration the kinds of insects in which the public seems to be most interested. In his 33 years with county agricultural commissioner's offices, the insects most frequently brought into the office for identification have been almost exclusively common, large, distinctive beetles, butterflies, and moths. As a consequence, the number of species in this field guide is weighted in favor of a few orders, particularly Coleoptera (beetles) and Lepidoptera (butterflies and moths).

One criterion he did not consider as part of the basis for inclusion in this book was the economic or social value of an insect. Much literature in the past has described insects in terms of how they affect humans, either in a positive way because they prey on other insects, or in a negative way because they do harm to crops or the lives of humans or domesticated animals or have been simply seen as pests. Our intent is to present insects for their intrinsic value, as players in natural processes, without the good/bad label.

Pete has underrepresented or omitted several orders of insects, even some that are very common, for a variety of reasons. Some insect groups likely to be encountered in daily life, such as mosquitoes, termites, or ants, are very difficult to identify to species, and many of the species that are small in size cannot be adequately identified by photograph. Many introduced insects that are common, such as honey bees, are not covered because this book emphasizes native species. Some aquatic insects that are well known to anglers, such as caddisflies and mayflies, have been omitted because they are well covered in several books on flyfishing and in the scientific literature; such insects are important to salmon and trout fisheries, but no more important than, say, dragonflies, which are included in this guide. Few household and garden "pests" are dealt with in this book because information about these species is often presented in cooperative extension pamphlets and books on pest control. The economic impacts of insects considered to be agricultural pests are discussed in other books that specifically deal with these species as pests; many of these pest species are nonnative (they have been introduced from Europe and Asia), and, again, this is not a book about pests.

The focus of this guide is the identification of insects (and some non-insect invertebrates). In order to make this guide easy to use, especially for the non-scientist, Pete has used photographs as the basis for the format, with picture-based keys. All the photographs in this field guide were taken with a single lens reflex (SLR) camera, usually handheld with no flash. The slide film used was ISO 50 or 100. All but a few of the photographs are of live insects and taken in the field.

Also for the sake of simplicity, we have attempted to keep the entomological jargon to a minimum. The few entomological terms that we determined needed definition are located in the glossary. However, unlike other field guides that primarily use common names for insects (and non-insect invertebrates), this guide is based on scientific names. We have emphasized scientific names because, as stated previously, they are standardized and more reliable than common names, and many of the insect species in this book do not have a common name or names.

Even so, the scientific name, as well as the classification, of a species may be different depending on which source the reader uses. Name changes can be very frustrating, and perhaps discouraging, to the non-scientist. We have tried to present the most recent classification and the most recent and most commonly used scientific and common names; for the most part, the classification and names used in this field guide are based on *Nomina Insecta Nearctica: A Check List of the Insects of North America* edited by R. W. Poole and P. Gentili.

The identification portion of this book is separated into two parts, one on the insects and the other on the non-insect terrestrial invertebrates. Within these sections, the groups of species are organized alphabetically, first by order, then by families within the order, and lastly by species within the family. Thus the arrangement of species presented here does not follow a hierarchical type of classification system. The orders are preceded by a picture key to the orders, and each order composed of more than a few families has a picture key to the families within that order. (Note: the size of a family or genus refers to the number of species within that family or genus.)

The account for each species is accompanied by a photograph of one or more of its life (developmental) stages and/or a structure (for example, plant gall) associated with it. For many of the species, particularly those of moths and butterflies (order Lepidoptera) and stink bugs (family Pentatomidae, order Hemiptera), Pete has included the immature stage(s) because, in most cases, one is just as likely to find this form of the species as one would the adult; a case in point is moth larvae, which are more likely to be seen than the adults since the adults of many moth species fly at night. Also, during certain times of the year, one can find the larva but not the adult.

Each species account is organized as follows. First is the species name, that is, the scientific name for the species. Although we have provided the most current scientific name, be aware that the name is not written in stone and hence may be different in other literature. A common name(s) may also be listed under the species name: a species may have multiple common names or none at all. The common name we give is generally the most widely used English name for the species in the literature.

Next follows a description of the distinguishing field marks of one or more life stages of the invertebrate, or, in the case of insect species that induce the formation of galls, the structure of the galls. The description is generally limited to those characteristics that we feel will help the reader identify the insect. If a life stage is not listed, it may be for one of the following reasons: (1) it is the stage least likely to be encountered in the field, (2) there may be little difference between adult and nymph, or (3) we did not find any information on that stage. This field guide uses the common practice found in other insect field guides, particularly those on butterflies and moths, of indicating the dorsal and ventral sides of an insect's body or wings as "above" and "below," respectively.

Body length and wingspan are the measurements commonly used to indicate the size of an invertebrate. Body length, used for invertebrates other than adult moths and butterflies, is the linear measurement in millimeters from the front of the animal's head (not including antennae) to the tip of its abdomen (not including terminal appendages). This measurement applies to the adult stage of the animal unless otherwise noted. For adult moths and butterflies, this book employs the common practice found in other field guides of using wingspan to compare the sizes of the different species of moths and butterflies. Wingspan is the linear measurement in millimeters from tip to tip of the outstretched forewings. This section may include the length, width, or height of galls.

The food source of one or more life stages (for example, adult, larva) of the species is given next. If a stage is not listed, we did not find any information regarding the food source or the animal does not feed at that stage. The common name (if any) and scientific name of host plants are listed.

The "found" section presents the general distribution of the species in the Pacific Northwest region (and may also indicate distribution outside the region). This part of the account may also include microhabitat information. Additional information about the species, including other scientific or common names of the species that might be encountered in other books, may be provided at the end of the account.

Searching for Insects

Insects can be found almost everywhere; however, certain habitats are more likely to yield a larger number and a wider variety of insects. The best places to look are on plants, where most insects will probably be found on the flowers and leaves; others may be on or in the stem, bark, wood, roots, or inside galls or the fruits or seeds of flowering plants.

Some species of insects can be found on the ground, in leaf litter, under the bark of dead or dying trees or logs, or under debris such as stones and boards (if these objects are moved, they should be carefully put back where they were found). Other species can be found on or in fungi, dung, or decaying plant or animal material. Many insect species live in or near water, either their entire lives or only during certain stages of their lives.

Although different insect species are active at different times of the year, the best time to look for the adults is usually early spring until late fall, when most plants are in bloom. Pete tries to time his search to coincide with where the flowering season is at its peak. He starts by looking for insects in late winter/early spring in areas that warm up first, which are usually near rivers or other bodies of water in inland valleys. Next, he goes to coastal areas, and then, as the season progresses, to higher elevations. After the flowering season peaks and summer turns into fall, insects can still be found in riparian areas and in other places that have late-flowering plants or lush foliage.

Different species can also be found at different times of the day, but flying insects are usually more active during the warmer periods of the day. Some species of flying insects, such as moths and lamellicorn beetles, come out during warm nights.

In order to identify insects, amateur as well as professional entomologists have traditionally captured them to be preserved as specimens in collections. Although collecting insects is the best way to identify them, it is becoming more common to observe and identify insects in their natural habitats while disturbing them as little as possible. To this end, close-focusing binoculars are an essential tool. This type of binoculars is capable of focusing quickly and on objects less than 6 feet away. A pair of 7×35 binoculars is usually adequate for most field observations.

Another piece of equipment that is excellent for use in identifying insects is the camera. Although preserving a specimen is the best way to have a permanent record of a species, taking a photograph of an insect is not only an accurate way of identifying most species but also a way of documenting a species without directly impacting it. Although Pete's camera of choice has been a single lens reflex (SLR) camera with a macro lens, he now considers a digital camera to be better. Digital cameras are lighter in weight and cheaper to purchase and to maintain; they do not need film, and the photographs can be immediately evaluated as to quality and then directly downloaded to a computer.

Good fieldwork requires additional tools. Those Pete carries into the field include a knife for prying up bark or digging into wood, soil, or other such media in which insects may hide; a hand lens for examining insects close-up; plant as well as insect field guides; and a notebook for recording such basic items as species name, host plant(s), date, and locality (he also may include such data as weather conditions, type of habitat, and insect behavior). Also handy to have are a small metric ruler and a pair of forceps, for manipulating dead insects or ones that may sting or bite.

If Pete has to capture an insect in order to identify it, he usually does so by hand. In addition to this method, there are several pieces of equipment one can use. One item that the public usually associates with entomologists is what is commonly referred to as the "butterfly net." This type of net, with the more accurate name of *aerial net*, is not only used to catch butterflies but other flying insects as well. Aerial nets are usually lightweight, durable, and have a bag made of fine-meshed material that is easy to see through. To capture non-flying insects, another type of net, called a *sweep net*, is used to beat or brush through vegetation. These nets have a bag made of canvas or heavy muslin. Whether caught by hand or by net, insects should be handled gently and with respect.

If an insect needs to be collected, but kept alive, to confirm its identification, it should be transported in some kind of a container: the best are made of hard plastic—Pete uses empty film containers for most insects. It is very important to transport the field containers in an ice chest (with ice), especially in hot weather, because the cold causes the insect to become torpid and, consequently, it is less likely to be injured. More detailed information about the methods and equipment used for collecting and preserving specimens is readily available in other sources.

Insects and Native Plants

Most of the insects in this field guide are native to the Pacific Northwest, and, as a result, most depend on native plants directly or indirectly for their survival. Increasingly, areas of native vegetation are being consumed by urban and suburban landscapes, where usually most, if not all, the native vegetation has been replaced with buildings, asphalt, and ornamental (that is, non-native) plants. Consequently, native wildlife populations, including those of insects, have been greatly reduced or extirpated in these areas.

There are efforts to reverse or at least mitigate this trend, and there is a growing awareness of how important a part the native insect fauna plays in these native landscapes. Since increased urbanization is a foregone conclusion, efforts to promote the use of native plants in gardening and landscaping urban and suburban areas will play a much larger role in the conservation of native plants and consequently of native wildlife populations.

Native plants in a perennial garden provide the stable environment needed to maintain healthy populations of insects, which in turn help maintain the health of the native plants. Providing this kind of environment will not only attract more insects and a wider variety of insect species to one's garden, it will also attract other wildlife that feed on these insects. After landscaping our property with native plants, we have seen a great increase in the number and diversity of animals, especially birds, on the property.

Even one native plant can make a difference. For example, one willow tree can provide food and shelter for many species of insects, such as aphids, ladybird beetles, predaceous stink bugs, and the larvae of leaf beetles, western tiger swallowtails and Lorquin's admirals. In addition, as a plant ages, the composition of insect species, as well as other wildlife species, associated with the plant changes over time. A young willow attracts leaf feeders like lepidoptera larvae and sucking insects like aphids. As the willow matures, its blooms provide pollen and nectar for pollinators like bumble bees, flies, and butterflies; its larger biomass in the form of leaves provides sustenance for insects such as tent caterpillars, leaf beetles, and galls; and insects such as bark beetles and boring beetles feed under the bark. As the health of the mature willow declines, other insect species feed on the tree's rotting wood and the associated fruiting bodies of fungi. As other wildlife are attracted to the plant at different stages of its life, they in turn create niches for different species of insects; for example, in older trees, sapsuckers drill holes in the tree bark to access the sap that oozes from these holes, and the oozing sap, in turn, is a food source for insects like ants, flies, true bugs, and moths.

Cultivating a variety of plants is even better because in doing so one creates an even more complex environment with more microhabitats for wildlife. Each plant in its own way makes a unique contribution to the diversity of the garden. There is also an interaction between the different plants that creates a synergy involving enhancement of the soil, complexity in canopy cover, and buildup of invertebrates in the duff, all of which means more food and shelter for the insects and other wildlife.

Many people seem to think that encouraging more insects to visit one's garden will result in significant insect damage to the plants. In addition, most of us have been taught to have little tolerance for most insects in the garden—"the only good insect is a dead insect" (the volume of insecticides sold

Native plant garden featuring showy milkweed, woolly sunflower, and clarkia.

in garden shops is testimony to this philosophy). In reality, the more diverse the species of insects in a garden, the higher the probability that there will be a balance in predator-prey interactions, and, consequently, less chance that there will be significant damage to the plants: not only will there be insects eating plants, but there will also be insects eating other insects. We should look at the native insect fauna not as something that threatens the garden but rather as an essential part of the processes that make the garden what it is, and view interactions between native insects and plants as normal and healthy.

Not only is a landscape planted with native vegetation a boon for native insects, it has another benefit for humans: it allows us to maintain regular contact with "wildness," which is so important in instilling and maintaining an appreciation and even love of "nature."

The following are postal and/or e-mail addresses of the state native plant societies in the Pacific Northwest region, which can provide information on the specific plants native to where the reader lives:

Native Plant Society of British Columbia
information@npsbc.org

Washington Native Plant Society
7400 Sand Point Way NE
Seattle, WA 98115
wnps@wnps.org

Native Plant Society of Oregon
P.O. Box 902
Eugene, OR 97440-0902

California Native Plant Society
1722 J Street, Suite 17
Sacramento, CA 95814
cnps@cnps.org

In raising and photographing insects, Pete has gained a unique insight into and a great respect for these wonderful creatures, and, through Pete, Judy has also gained a better understanding of them. We hope this book will engender in the reader this same sense of awe and respect and stimulate a desire to learn more about insects as well as the other animals and plants and natural communities that represent the diversity of life.

PART ONE
INSECT ORDERS

Key to Insect Orders

BEETLES
Order Coleoptera, page 24

FLIES
Order Diptera, page 86

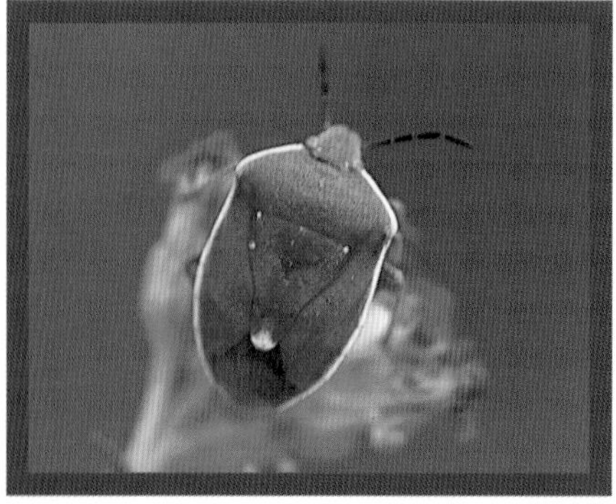

BUGS
Order Hemiptera, page 94

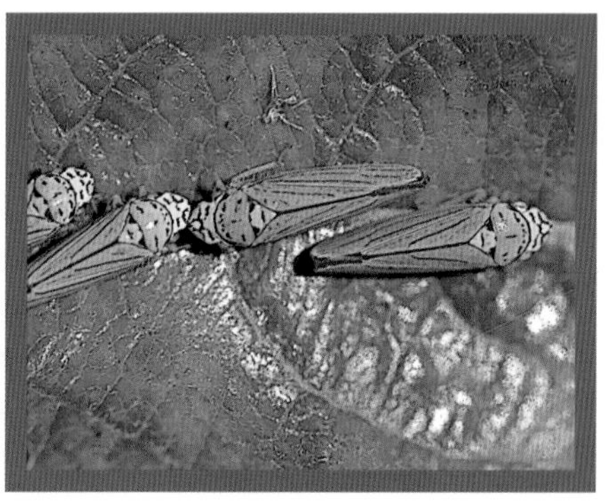

CICADAS, LEAFHOPPERS & ALLIES
Order Homoptera, page 108

WASPS, ANTS & BEES
Order Hymenoptera, page 118

BUTTERFLIES & MOTHS
Order Lepidoptera, page 136

ANTLIONS, SNAKEFLIES & ALLIES
Order Neuroptera, page 242

DRAGONFLIES & DAMSELFLIES
Order Odonata, page 245

GRASSHOPPERS, CRICKETS & KATYDIDS
Order Orthoptera, page 255

STONEFLIES
Order Plecoptera, page 260

Coleoptera (Greek, *coleo*, sheath; *ptera*, wings) is the largest order in the animal kingdom and, with approximately 290,000 species described, contains an estimated 37 percent of all known insect species. More than 23,700 species have been recorded in the United States and Canada. Approximately one-third of the total number of insect species found in the Pacific Northwest are beetles. This order includes some of the largest and smallest insects. Beetles vary considerably in habits and can be found almost everywhere.

Beetles are easily distinguishable from other insect orders by the structure of their wings. Most species have two pairs of wings, with the forewings modified into hard or leathery sheathlike wingcovers called *elytra* (hence the order name). The hindwings are membranous and usually longer than the forewings; when the beetle is at rest, they are folded under the forewings, allowing the beetle to move easily when not flying. The hindwings are the only wings used for flight. The modification of the forewings into shields that protect the beetle's abdomen and the ability of the flying wings (hindwings) to be folded under these wingcovers are two probable reasons for the success of this order.

The adult body, which is also usually hard or leathery, occurs in a diversity of forms. The majority of species are black or dark brown, but others are brightly colored. The antennae are of various shapes and are helpful as an identification tool. The mouthparts are adapted for biting and chewing, and the jaws are well developed.

Beetles undergo complete metamorphosis. Although the larvae are commonly called *grubs*, which suggests a maggotlike shape, they vary in form. Most species have large heads, and all have chewing mouthparts. Some species are legless.

Beetles occupy all terrestrial and freshwater habitats, and the mouthparts of both the adults and larvae enable them to feed on a great variety of plants or animals. Many species are scavengers; a few are parasites.

Some species are regarded as pests because they feed on crop or garden plants and stored plant and animal products, but most beetles are considered to be beneficial. Many species eat other insects that are plant pests, while other species pollinate flowers. A very large number of species in families that are scavengers feed on feces and decaying plant and animal material and consequently are extremely important in recycling organic matter into nutrients other organisms can use.

Key to Coleoptera Families

METALLIC WOOD-BORING BEETLES
Family Buprestidae, page 29

GROUND BEETLES
Family Carabidae, page 31

LONG-HORNED BEETLES
Family Cerambycidae, page 33

LEAF BEETLES
Family Chrysomelidae, page 46

TIGER BEETLES
Family Cicindelidae, page 54

CHECKERED BEETLES
Family Cleridae, page 55

Key to Coleoptera Families, continued

LADYBIRD BEETLES
Family Coccinellidae, page 57

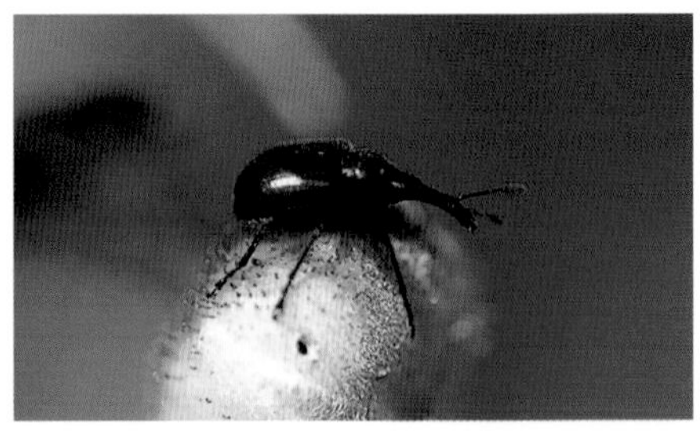

SNOUT BEETLES
Family Curculionidae, page 66

PREDACEOUS DIVING BEETLES
Family Dytiscidae, page 67

CLICK BEETLES
Family Elateridae, page 68

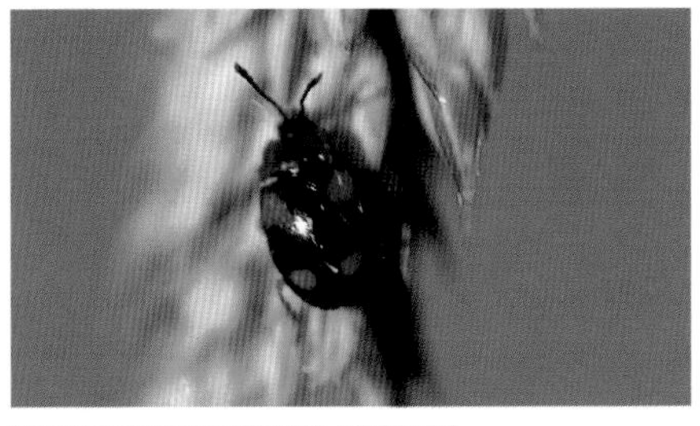

HANDSOME FUNGUS BEETLES
Family Endomychidae, page 69

MARSH BEETLES
Family Helodidae, page 70

CLOWN BEETLES
Family Histeridae, page 70

FIREFLY BEETLES
Family Lampyridae, page 71

STAG BEETLES
Family Lucanidae, page 73

NET-WINGED BEETLES
Family Lycidae, page 74

BLISTER BEETLES
Family Meloidae, page 74

SOFT-WINGED FLOWER BEETLES
Family Melyridae, page 75

POLLEN-FEEDING BEETLES
Family Oedemeridae, page 76

BARK-GNAWING BEETLES
Family Ostomidae, page 76

GLOWWORM BEETLES
Family Phengodidae, page 77

LAMELLICORN BEETLES
Family Scarabaeidae, page 78

Key to Coleoptera Families, continued

BARK BEETLES
Family Scolytidae, page 81

CARRION BEETLES
Family Silphidae, page 82

ROVE BEETLES
Family Staphylinidae, page 83

DARKLING BEETLES
Family Tenebrionidae, page 84

IRONCLAD BEETLES
Family Zopheridae, page 85

This large family gets part of its common name from the distinctive metallic sheen of the adults. These mostly medium-sized beetles have hard, cylindrical to flattened oval bodies. They are often found on the bark of dead and dying trees, although some species frequent flowers, where they feed on pollen. When disturbed, they quickly take flight or drop to the ground. The larvae are elongate and flattened. Most species have their anterior end broadened and flattened, in contrast to the long-horned beetles (Cerambycidae), and are often known as *flat-headed borers.* They feed under the bark in the cambium layer or in the heartwood of trees, and the holes they bore into the wood are oval in shape.

Acmaeodera connexa

SPOTTED FLOWER BUPRESTID

ADULT Wingcovers black with light brown to white markings. **BODY LENGTH** 12 mm. **FOOD** Larva: Wood (borer). **FOUND** Throughout the region. Look for adults on flowers in the spring.

This species is one of the most common buprestids.

Acmaeodera connexa

Agrilus politus

WILLOW TWIG GIRDLER

ADULT Slender, dark bronzy green to black; wingcovers wider immediately posterior to midline. **BODY LENGTH** 8 mm. **FOOD** Adult: Black cottonwood (*Populus balsamifera*), willow (*Salix* spp.) (leaves). Larva: Wood (borer). **FOUND** Throughout the region. Look for adults on leaves and stems of host plants from late spring (most common) into summer.

Agrilus politus

Buprestis aurulenta

Buprestis aurulenta

GOLDEN BUPRESTID

ADULT Iridescent green; wingcovers with coppery red margins and each with five (appears to be four) longitudinal grooves. **BODY LENGTH** 17 mm. **FOOD** Larva: Wood (borer). **FOUND** From British Columbia to California.

The adult retains its iridescence long after it is dead.

Buprestis langi, female and male

Buprestis langi

LANG'S BUPRESTID

ADULT Iridescent green; wingcovers unspotted or each with up to three white spots. Males are smaller than females and usually spotted (as in photograph). **BODY LENGTH** 20 mm. **FOOD** Adult: Black cottonwood (*Populus balsamifera*) (leaves). Larva: Wood (borer). **FOUND** Throughout the region. Look for adults on host trees from late spring to early summer.

Chalcophora virginiensis

Chalcophora virginiensis

SCULPTURED PINE BORER

ADULT Black to dark brown with some bronze highlights; wingcovers and pronotum heavily sculpted. **BODY LENGTH** 30 mm. **LARVA** White; anterior end broadened and flattened. **FOOD** Larva: Wood (borer). **FOUND** Throughout the region.

Also known as *Chalcophora angulicollis.*

Chalcophora virginiensis, larva

■ GROUND BEETLES Family Carabidae

This is the second-largest family of beetles in North America. Although there is considerable variation in size, color, and shape within this family, most members have shiny black, elongate, somewhat flattened, medium- to large-sized bodies. The legs are long, and the pronotum is often narrower than the wingcovers. The adults usually hide under objects during the day and hunt for prey at night. When disturbed, they run rapidly on the ground, but seldom fly. Many give off an unpleasant smell when threatened. The adults and larvae of most species prey on other invertebrates. They are beneficial in the garden but can be very sensitive to human activity and insecticides.

Brachinus quadripennis

BOMBARDIER BEETLE

ADULT Wingcovers iridescent blue; pronotum reddish brown. **BODY LENGTH** 12 mm. **FOOD** Adult: Invertebrates. Larva: (parasitic). **FOUND** Throughout the region, under rocks on the banks of streams and rivers.

When threatened, the adult may eject a liquid from glands at the rear of its abdomen. The ejected liquid vaporizes immediately on contact with the air, causing a popping sound; it can burn tender human skin.

Brachinus quadripennis

Carabus granulatus

ADULT Black; wingcovers each with three rows of bumps. **BODY LENGTH** 20 mm. **FOOD** Adult: Invertebrates. **FOUND** Throughout the region, most commonly in conifer forests, on the ground or under bark.

Carabus granulatus

Carabus nemoralis

EUROPEAN GROUND BEETLE

ADULT Black with purple iridescence; wingcovers each with three rows of pits. **BODY LENGTH** 22 mm. **FOOD** Adult: Invertebrates. **FOUND** Throughout much of the northern part of the U.S., commonly in gardens.

This species was introduced from Europe for biological control. Gardeners consider it a beneficial insect since it feeds on garden pests.

Carabus nemoralis

Elaphrus purpurans (bottom) and *Opisthius richardsoni*

Elaphrus purpurans

ADULT Dark brown, often with purplish green iridescence; pronotum covered with large shallow pits; wingcovers covered with large shallow pits and shiny black patches. **BODY LENGTH** 6 mm. **FOOD** Adult and larva: Invertebrates. **FOUND** Throughout the region. Look for adults in muddy or sandy areas along streams, rivers, and ponds.

Also known as *Elaphrus pallipes*. This species often occurs with tiger beetles (*Cicindela* spp.) and *Opisthius richardsoni*. Its movements are exactly like those of tiger beetles, which run rapidly along the ground looking for prey. They are also very sensitive to movement, making them very difficult to approach, and therefore especially difficult to photograph.

Omophron ovale

Omophron ovale

SAVAGE BEETLE

ADULT Ovate; dark green with pale yellow markings. **BODY LENGTH** 7 mm. **FOOD** Adult and larva: Invertebrates. **FOUND** Throughout the region. Adults most common under rocks near flowing water.

During the day, the adult hollows out an area slightly bigger than itself in the sand or soil under a rock or piece of wood. It emerges from this protective cell at night to hunt.

Opisthius richardsoni

ADULT Dark brown; wingcovers covered with large shallow pits. **BODY LENGTH** 8 mm. **FOOD** Adult and larva: Invertebrates. **FOUND** Throughout the region. Look for adults in muddy or sandy areas along streams, rivers, and ponds.

Also known as *Elaphrus americanus*. This species often occurs with tiger beetles (*Cicindela* spp.) and *E. purpurans*. Its movements are exactly like those of tiger beetles, which run rapidly along the ground looking for prey. They are also very sensitive to movement, making them very difficult to approach, and therefore especially difficult to photograph.

Promecognathus laevissimus

ADULT Smooth, shiny black; jaws very large. **BODY LENGTH** 16 mm. **FOOD** Adult: Millipedes. **FOUND** Throughout the region. Adults hide under wood, tree bark, or rocks during the day.

The adult has specialized jaws to prey on millipedes.

Promecognathus laevissimus

■ LONG-HORNED BEETLES Family Cerambycidae

This very large and well-known family of wood-boring beetles is well documented in the scientific literature: see Linsley and Chemsak (1997) and Hatch (1957–1971); Hatch is particularly useful to the amateur entomologist because one can usually skip the identification key and go directly to his excellent black-and-white drawings. These beetles are commonly found in forested areas throughout the Pacific Northwest. The adults are medium to large in size and in most species have elongate, cylindrical bodies and long antennae (hence their common name). Many are distinctly colored and patterned. The adults feed on wood, leaves, roots, pollen, and, rarely, other insects. Many of the more brightly colored adults are found on flowers during the day; in particular, those in the subfamily Lepurinae are attracted to flowers, making them easy to find in the spring and early summer (many of the long-horned beetle species in this field guide are in this subfamily). When on flowers, the beetles are so intent on eating pollen or mating, or both, that the observer can get very close to the preoccupied beetle without causing it to fly away or drop to the ground. Others, usually not brightly colored, emerge from under logs and loose bark only at night. Most long-horns are wood-boring in the larval stage. The larvae emerge from eggs that have been laid by the adults in crevices in the bark. They are elongate, cylindrical, white, and appear legless. Their anterior end is not broadened and flattened, as in the metallic wood-boring beetles (Buprestidae), and they are often called *round-headed borers.* The holes they make in the wood are circular in shape. A few species attack living trees, but most seem to prefer weakened and dying trees or freshly cut logs.

Anastrangalia laetifica, female and male

Anastrangalia laetifica

DIMORPHIC LONG-HORNED BEETLE

ADULT Female, wingcovers red, each with two black spots; male, usually all black, or less commonly brown with black pattern. **BODY LENGTH** 13 mm. **FOOD** Larva: Wood (borer). **FOUND** Throughout the region. Adult found spring through midsummer usually as mating pairs on flowers.

This species is one of the most common long-horned beetles.

Anastrangalia sanguinea

Anastrangalia sanguinea

ADULT Mostly black; wingcovers brown to reddish brown in male and reddish brown in female. **BODY LENGTH** 13 mm. **FOOD** Larva: Wood (borer). **FOUND** Throughout the region.

Brachyleptura vexatrix

Brachyleptura vexatrix

ADULT Wingcovers black or black with varying amounts of yellow, each usually with medial black spot or band. **BODY LENGTH** 13 mm. **FOOD** Larva: Wood (borer). **FOUND** Throughout the region. Adult found on flowers.

Brachysomida californicus

ADULT Wingcovers black, brown, iridescent blue, or green. **BODY LENGTH** 11 mm. **FOOD** Larva: Wood (borer). **FOUND** Throughout the region. Adult is very common on flowers.

Brachysomida californicus

Brachysomida californicus

This species is difficult to identify; Linsley and Chemsak (1997) list 43 synonyms. Its small size and solid color are helpful in separating it from similar species.

Callimoxys sanguinicollis, male and female

Callimoxys sanguinicollis

ADULT Wingcovers black and acutely tapered toward tip; wings mostly exposed; pronotum red in females, black in males; femurs club-shaped with swollen part of club black. **BODY LENGTH** 12 mm. **FOOD** Larva: Wood (borer). **FOUND** Throughout the U.S. Adult often found in great numbers feeding and mating on California-lilac (*Ceanothus* spp.) flowers.

Centrodera spurcus

Centrodera spurcus

YELLOW DOUGLAS-FIR BORER

ADULT Wingcovers light reddish brown, each with small dark brown spot on outer margin; pronotum with tooth on each side; eyes black. **BODY LENGTH** 30 mm. **FOOD** Larva: Wood (borer). **FOUND** Throughout the region. Adult attracted to lights at night.

Cosmosalia chrysocoma

Cosmosalia chrysocoma

YELLOW VELVET LONG-HORNED BEETLE

ADULT Wingcovers usually pale brown, densely covered with golden yellow hairs; pronotum black, covered with golden yellow hairs. **BODY LENGTH** 18 mm. **FOOD** Larva: Wood (borer). **FOUND** Common every year throughout the region. Look for adult on wild buckwheat (*Eriogonum* spp.) and California-lilac (*Ceanothus* spp.) flowers.

Desmocerus auripennis

Desmocerus auripennis

ELDERBERRY LONG-HORNED BEETLE

ADULT Extremely variable in size and pattern; wingcovers green with red border (border often much wider than it appears in photographs of this species; in pinned specimens, red becomes pale yellow). **BODY LENGTH** 16 mm. **FOOD** Larva: Elderberry (*Sambucus* spp.) (wood borer). **FOUND** Throughout the region.

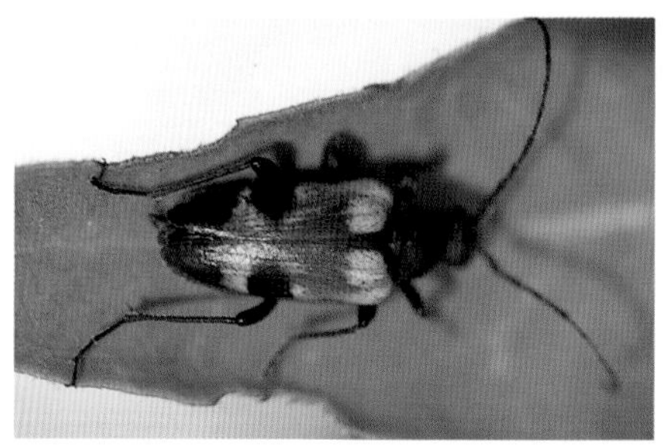

Dorcasina matthewsii

Dorcasina matthewsii

ADULT Wingcovers light brown with black tip and each with black spot in middle. **BODY LENGTH** 20 mm. **FOOD** Larva: Wood (borer). **FOUND** Throughout the region.

Ergates spiculatus

Ergates spiculatus

SPINED WOODBORER

ADULT Brown; pronotum with many small spines on margin. May be confused with *Prionus californicus*, but pronotum of latter has three large spines. **BODY LENGTH** 65 mm. **FOOD** Larva: Wood (borer). **FOUND** Throughout the region. Adult often attracted to lights at night.

This species and *Prionus californicus* are the largest beetles in the region.

Judolia instabilis, typical pattern

Judolia instabilis

INSTABLE LONG-HORNED BEETLE

ADULT Black; wingcovers with creamy white areas that are extremely variable (see Hatch [1957–1971] for many of the variations). **BODY LENGTH** 12 mm. **FOOD** Larva: Wood (borer). **FOUND** Throughout the region. Adult found on flowers.

Lampropterus cyanipenne

ADULT Pronotum red in female, black in male; wingcovers dark iridescent green or dark iridescent blue; legs red with black at joints. **BODY LENGTH** 8 mm. **FOOD** Larva: Wood (borer). **FOUND** Oregon and California.

Lampropterus cyanipenne, female

Leptalia macilenta

ADULT Wingcovers dark brown, each with broad, light brown longitudinal line down center; male may have completely light brown wingcovers. **BODY LENGTH** 11 mm. **FOOD** Larva: Wood (borer). **FOUND** Throughout the region. Adult found on flowers.

Also known as *Leptalia frankenhausen*. This species can be easily identified by size and wingcover pattern.

Leptalia macilenta

Leptura obliterata

ADULT Wingcovers pale yellow with black areas in two basic patterns—wingcover tip black and median area with incomplete black band (as shown), or shoulders with black stripes or spots; pronotum black with pale yellow margin. Pronotum pattern is very consistent and should be used for identification since size and wingcover pattern are so variable in this species. **BODY LENGTH** 18 mm. **FOOD** Larva: Wood (borer). **FOUND** Throughout the region. Adult found on flowers.

This species is one of the most common large long-horned beetles in the region.

Leptura obliterata

Lepturopsis dolorosa

ADULT Black; pronotum with longitudinal groove. **BODY LENGTH** 15 mm. **FOOD** Larva: Wood (borer). **FOUND** Throughout the region. Adult found on flowers.

Lepturopsis dolorosa

Molorchus longicollis

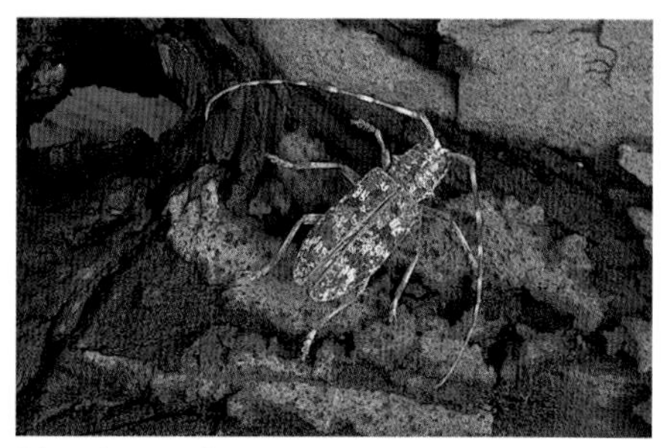

Monochamus obtusus

Monochamus scutellatus

Necydalis cavipennis

Molorchus longicollis

ADULT Wasp mimic—black to reddish black; wingcovers light red to yellow and half the length of abdomen. **BODY LENGTH** 7 mm. **FOOD** Adult: Plants (pollen). Larva: Wood (borer). **FOUND** Throughout the region. Adult common on California-lilac (*Ceanothus* spp.) flowers in spring.

Monochamus obtusus

OBTUSE SAWYER

ADULT Brown with white patches; pronotum with tooth on each side; antennae banded brown and white. **BODY LENGTH** 20 mm. **FOOD** Larva: Wood (borer). **FOUND** Throughout the region.

Monochamus scutellatus

BLACK PINE SAWYER

ADULT Black; wingcovers each with small white area at base; pronotum with tooth on each side; male antennae may be twice length of body. **BODY LENGTH** 26 mm. **FOOD** Larva: Wood (borer). **FOUND** Throughout the region.

Necydalis cavipennis

ADULT Wingcovers coppery brown and only one-fourth length of wings; expanded parts of legs coppery brown otherwise black. Very similar to *Necydalis laevicolis* but longer and more robust. **BODY LENGTH** 23 mm. **FOOD** Larva: Wood (borer). **FOUND** Throughout the region.

Neoalosterna rubida

ADULT Wingcovers orange to reddish orange with narrow black margin at tip; pronotum black. **BODY LENGTH** 14 mm. **FOOD** Larva: Wood (borer). **FOUND** Throughout the region.

Neoalosterna rubida

Neobellamira delicata

ADULT Wingcovers taper acutely from base to posterior, light brown, each with black line extending from base to tip; head and pronotum black; abdomen and legs red. Wingcover shape and color sets this species apart from other long-horned beetles. **BODY LENGTH** 15 mm. **FOOD** Larva: Wood (borer). **FOUND** Oregon and California, commonly on flowers.

Neobellamira delicata

Neoclytus balteatus

ADULT Male, wingcovers reddish brown (common) or black, each with three yellow or pale yellow bands. Female, wingcovers black, each with three yellow bands; pronotum with three yellow bands. **BODY LENGTH** 16 mm. **FOOD** Larva: Wood (borer). **FOUND** Throughout the region. Pete has commonly found adult males on the stems and leaves of the California-lilac *Ceanothus velutinus.*

Neoclytus balteatus, males

Neoclytus conjunctus

ADULT Wingcovers black with white areas and often with incomplete, white circle on each shoulder. **BODY LENGTH** 18 mm. **FOOD** Larva: Wood (borer). **FOUND** Throughout the region. Commonly emerges from firewood and other downed wood in late winter and spring.

Neoclytus conjunctus

Oberea schaumii

Oberea schaumii

TWIG BORER

ADULT Wingcovers, head, and antennae black; pronotum pale red with four raised black spots and one smaller central black spot. **BODY LENGTH** 15 mm. **FOOD** Larva: Willow (*Salix* spp.) and poplar (*Populus* spp.) (borer). **FOUND** Throughout the region. Look for adults on willow and poplar trees.

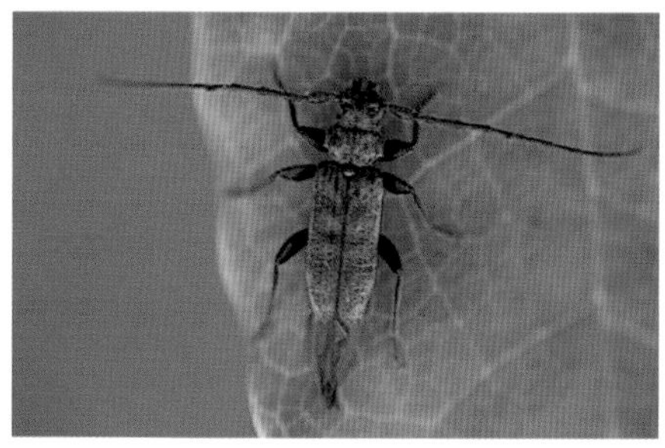

Opsimus quadrilineatus

Opsimus quadrilineatus

SPRUCE LIMB BORER

ADULT Brown; wingcovers covered with brown pubescence; pronotum covered with brown pubescence and with spine on each side; antennae at least as long as body. **BODY LENGTH** 12 mm. **FOOD** Larva: Wood (borer). **FOUND** Throughout the region. Pete has found adults emerging from wood beams inside a house.

Ortholeptura valida

Ortholeptura valida

ADULT Wingcovers pale brown, each with three black spots. **BODY LENGTH** 24 mm. **FOOD** Larva: Wood (borer). **FOUND** Throughout the region.

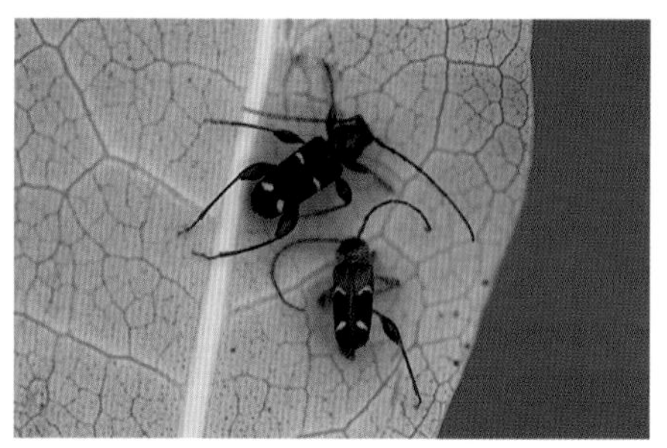

Phymatodes decussatum

Phymatodes decussatum

ADULT Variable in coloration and pattern—two most common patterns are (1) black with two white bands on each wingcover and (2) reddish brown with posterior three-quarters of wingcover black and each wingcover with two white bands. **BODY LENGTH** 9 mm. **FOOD** Larva: Wood (borer). **FOUND** Throughout the region. Larvae often brought into houses on firewood. In spring adults emerge from the firewood and usually end up at windows looking for a way to escape from the house.

Phymatodes infuscatum

ADULT Wingcovers black to dark brown with reddish brown base; body, legs, and antennae reddish brown. **BODY LENGTH** 16 mm. **FOOD** Larva: Wood (borer). **FOUND** Oregon and California. Pete has found adults emerging from oak firewood in the spring.

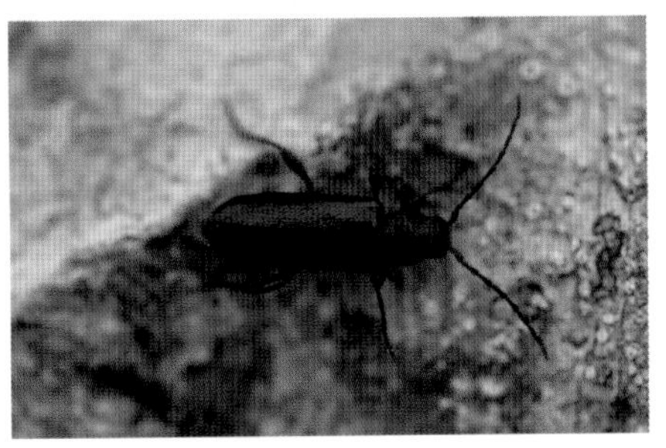

Phymatodes infuscatum

Pidonia gnathoides

ADULT Wingcovers reddish brown, each with black spot in middle. **BODY LENGTH** 9 mm. **FOOD** Larva: Wood (borer). **FOUND** Common throughout the region. Look for adult on flowers.

Pidonia gnathoides

Pidonia scripta

ADULT Wingcovers light brown and variably patterned with black spots and lines; pronotum, head, and legs red. **BODY LENGTH** 9 mm. **FOOD** Larva: Wood (borer). **FOUND** Very common throughout the region. Adult attracted to flowers.

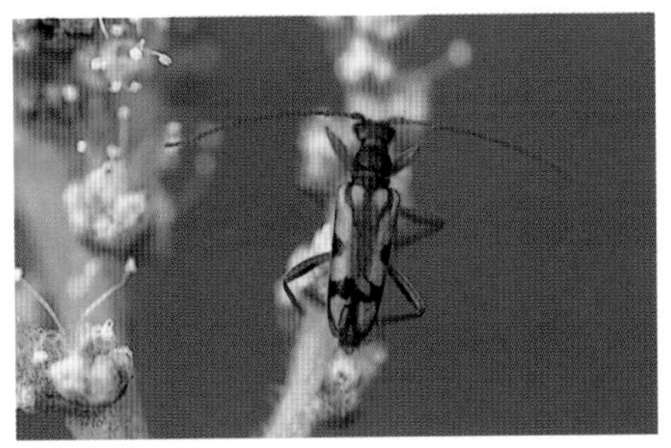

Pidonia scripta

Prionus californicus

CALIFORNIA PRIONUS

ADULT Wingcovers reddish brown to brown; pronotum reddish brown to brown with three large spines on each side; antennae long and robust. May be confused with *Ergates spiculatus*, but pronotum of latter has many small spines. **BODY LENGTH** 55 mm. **FOOD** Larva: Wood (borer). **FOUND** Throughout the region. Adult readily attracted to lights at night. Larva found in living roots of broadleaf trees or dead roots of broadleaf or coniferous trees.

The larva requires three to five years to mature.

Prionus californicus

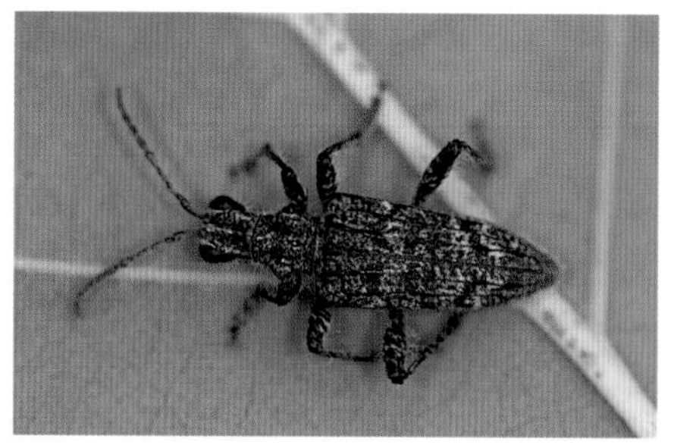

Rhagium inquisitor

Rhagium inquisitor

RIBBED PINE BORER

ADULT Wingcovers gray and brown and strongly ribbed; head and pronotum black and covered with gray pubescence. **BODY LENGTH** 18 mm. **FOOD** Larva: Wood (borer). **FOUND** Throughout most of the U.S. Look for adult on the trunks of fire-damaged or dead trees; it blends perfectly with tree bark and can only be seen when it moves.

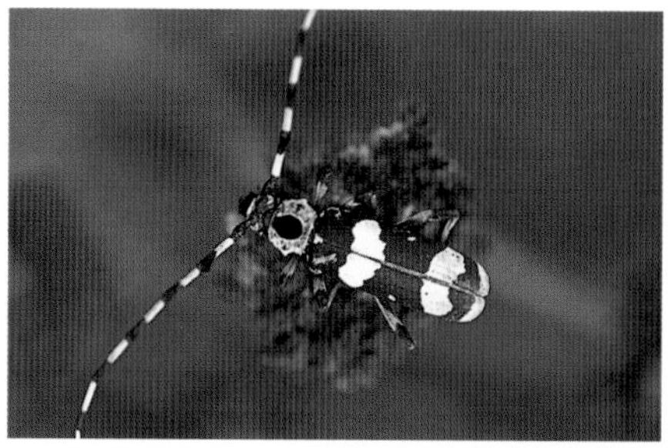

Rosalia funebra

Rosalia funebra

BANDED ALDER BORER

ADULT Wingcovers brown to dark brown, each with three white bands; pronotum pale white with large, black, oval spot; antennae banded white and black. **BODY LENGTH** 38 mm. **FOOD** Larva: Wood (borer). **FOUND** Throughout the region. Pete has found large numbers of adults in the spring on the bark of fallen alders.

Saperda populneus

Saperda populneus

ADULT Wingcovers dark brown to black with sparse golden yellow hairs; pronotum black with pair of brownish red lateral lines. **BODY LENGTH** 13 mm. **FOOD** Larva: Wood (borer). **FOUND** Throughout the U.S.

Stenocorus flavolineatus

Stenocorus flavolineatus

ADULT Wingcovers black, each with narrow, light brown, medial stripe. **BODY LENGTH** 25 mm. **FOOD** Larva: Wood (borer). **FOUND** Throughout the region.

Stenocorus vestitus

ADULT Wingcovers reddish brown; pronotum and head black with short golden hairs. **BODY LENGTH** 20 mm. **FOOD** Larva: Wood (borer). **FOUND** Throughout the region.

Stenostrophia amabilis

ADULT Wingcovers black, each with four yellow spots or bands. **BODY LENGTH** 14 mm. **FOOD** Larva: Wood (borer). **FOUND** Throughout the region. Adult found on flowers.

Stictoleptura canadensis

ADULT Wingcovers black with red base. **BODY LENGTH** 20 mm. **FOOD** Larva: Wood (borer). **FOUND** Throughout the U.S. Adult attracted to flowers.

Strophiona laeta

ADULT Wingcovers black, each with three golden yellow bands (may be connected) and pair of golden yellow apical spots. **BODY LENGTH** 16 mm. **FOOD** Larva: Wood (borer). **FOUND** Throughout the region.

Stenocorus vestitus

Stenostrophia amabilis

Stictoleptura canadensis

Strophiona laeta

Tetraopes basalis

Ulochaetes leoninus

Xestoleptura crassicornis

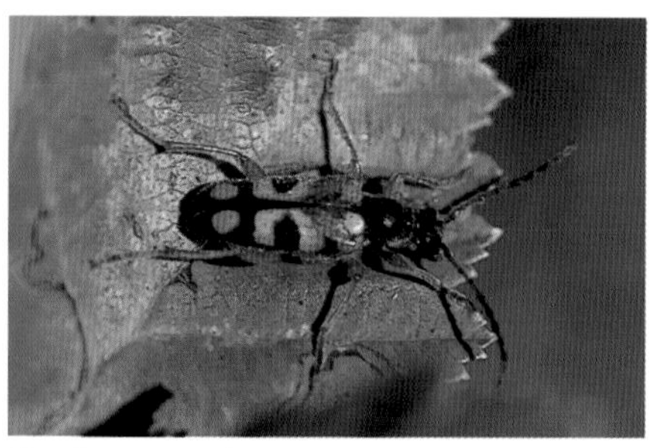

Xestoleptura crassipes

Tetraopes basalis

RED MILKWEED BORER

ADULT Robust; red with black spots. **BODY LENGTH** 17 mm. **FOOD** Adult and larva: Showy milkweed (*Asclepias speciosa*), narrowleaf milkweed (*A. fascicularis*), Kotolo milkweed (*A. eriocarpa*) (adult—flowers and leaves; larva—roots). **FOUND** Oregon and California (replaced by a similar species, *Tetraopes femoratus*, in British Columbia, Washington, northern Oregon, and eastern California).

The adult moves slowly and does not readily fly, even when disturbed by a pesky photographer.

Ulochaetes leoninus

LION BEETLE

ADULT Bumble bee mimic—black; wingcovers shortened and with yellow tip; thorax with yellow hairs; femurs with yellow band. **BODY LENGTH** 30 mm. **FOOD** Larva: Wood (borer). **FOUND** Throughout the region.

When handled, the female will try to "sting" with the tip of her abdomen.

Xestoleptura crassicornis

ADULT Wingcovers dark brown with golden yellow pubescence, each with three golden yellow bands and often with golden yellow spot near tip; pronotum, head, and legs reddish brown with golden yellow pubescence; eyes black. **BODY LENGTH** 17 mm. **FOOD** Larva: Wood (borer). **FOUND** Throughout the region. Adult attracted to flowers.

Xestoleptura crassipes

ADULT Wingcovers black with variably patterned, golden yellow areas and each with golden yellow subapical spot. **BODY LENGTH** 14 mm. **FOOD** Larva: Wood (borer). **FOUND** Throughout the region. Adult attracted to flowers.

Xylotrechus insignis

ADULT Male, wingcovers purplish brown with yellow pubescence, and yellow bands that are variable in size and usually faint. Female, wingcovers black with yellow bands; pronotum black with yellow anterior and posterior margins. Similar to *Xylotrechus nunenmacheri* except in latter, male with white pubescence and female pronotum with no yellow margins. **BODY LENGTH** 18 mm. **FOOD** Larva: Wood (borer). **FOUND** Oregon and California. Pete has found adults on willows (*Salix* spp.).

This species is more common than *Xylotrechus nunenmacheri.*

Xylotrechus insignis, male

Xylotrechus insignis, female

Xylotrechus longitarsis

ADULT Pronotum dark brown with white anterior margin; wingcovers dark brown with white bands, some incomplete. **BODY LENGTH** 20 mm. **FOOD** Larva: Wood (borer). **FOUND** Throughout the region. Adults on or under bark of Douglas-fir (*Pseudotsuga menziesii*).

Xylotrechus longitarsis

Xylotrechus nauticus

NAUTICAL BORER

ADULT Dark gray to dark brown; pubescent; wingcovers with three transverse irregular white lines. **BODY LENGTH** 16 mm. **FOOD** Larva: Wood (borer). **FOUND** Throughout the region. Adults commonly emerge from firewood stored inside buildings.

Xylotrechus nauticus

Xylotrechus nunenmacheri, male

Xylotrechus nunenmacheri

ADULT Male, wingcovers purplish brown with white pubescence and reduced, white bands that are usually faint; pronotum purplish brown with white pubescence. Female, wingcovers black with yellow bands; pronotum black. Similar to *Xylotrechus insignis* except in latter, male with yellow pubescence and female pronotum with yellow anterior and posterior margins. **BODY LENGTH** 18 mm. **FOOD** Larva: Wood (borer). **FOUND** Throughout the region; common except in California.

■ LEAF BEETLES Family Chrysomelidae

The members of this family are very common in the Pacific Northwest. The adults of most species are small (less than 13 mm in length), convex, oval, and brightly colored and distinctly patterned. Many are similar in appearance to ladybird beetles: both leaf beetles and ladybird beetles are often spotted and/or striped with black and red. The hind legs of many of the smaller species, called *flea beetles*, are enlarged for jumping. The adults and larvae feed on plants. Some are important agricultural and garden pests.

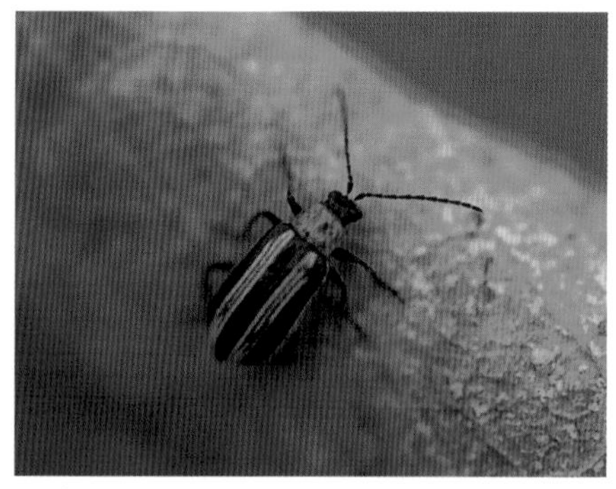

Acalymma trivittata

Acalymma trivittata

WESTERN STRIPED CUCUMBER BEETLE

ADULT Wingcovers pale white, each with one longitudinal black line and black interior margin; pronotum orange; head black. **BODY LENGTH** 6 mm. **FOOD** Adult: Plants, preferably herbaceous (leaves and flowers). Larva: Plants (roots). **FOUND** Throughout the region; very common in California.

This species may be a garden pest.

Altica ambiens

ALDER FLEA BEETLE

ADULT Iridescent dark blue. **BODY LENGTH** 6 mm. **LARVA** Dark brown or black. **FOOD** Adult and larva: Red alder (*Alnus rubra*) and other alders (*Alnus* spp.). **FOUND** Throughout much of the U.S.

Also known as *Macrohaltica ambiens*. Although this species is difficult to separate taxonomically from *Altica bimarginata*, we have included both in the field guide because they are always common and at times occur in staggering numbers. Hatch (1957–1971) uses sexual characteristics and a fold on the wingcovers to separate the two species. A more practical approach in the field is to use the host plant.

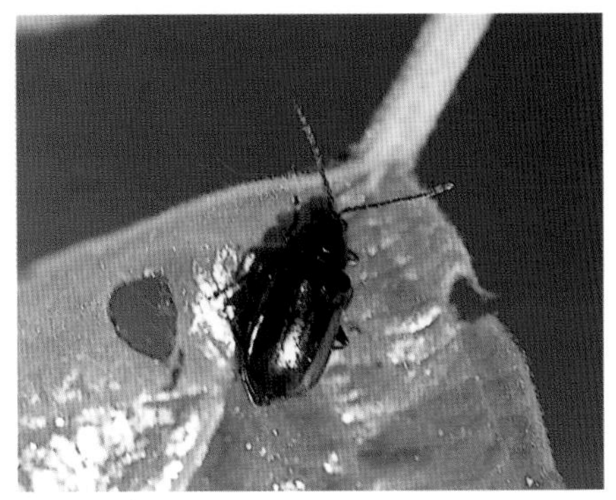

Altica ambiens

Altica bimarginata

WILLOW FLEA BEETLE

ADULT Iridescent blue. **BODY LENGTH** 6 mm. **LARVA** Dark brown to black. **FOOD** Adult and larva: Narrow-leaved willow (*Salix exigua*) and other willows (*Salix* spp.). **FOUND** Throughout the region.

Also known as *Macrohaltica bimarginata*. Adults form large mating congregations on or near host plants. See *Altica ambiens* for more information on identification.

Altica bimarginata

Calligrapha multipunctata

ADULT Wingcovers golden yellow with black spots and lines; pronotum black with golden yellow anterior margin. **BODY LENGTH** 8 mm. **LARVA** Pale green with orange head. **FOOD** Larva: Coastal willow (*Salix hookeriana*) and other willows (*Salix* spp.). **FOUND** From British Columbia to California; uncommon in California.

Calligrapha multipunctata

Chrysochus cobaltinus

Chrysolina quadrigemina

Chrysomela aeneicollis

Chrysomela aeneicollis

Chrysochus cobaltinus

COBALT LEAF BEETLE

ADULT Iridescent blue. **BODY LENGTH** 11 mm. **FOOD** Adult and larva: Bitter dogbane (*Apocynum androsaemifolium*) (adult—leaves; larva—roots). **FOUND** Throughout the region. Common wherever the host plant grows.

The adults are difficult to photograph because if they sense any movement, they immediately drop to the ground and hide. They produce an unpleasant smell when handled.

Chrysolina quadrigemina

KLAMATH WEED BEETLE

ADULT Wingcovers and pronotum iridescent blue, blue-green, or bronze with fine but distinct punctures. **BODY LENGTH** 6 mm. **FOOD** Adult and larva: Klamath weed (*Hypericum perforatum*) (adult—leaves; larva—roots). **FOUND** Throughout the region.

Also known as *Chrysolina gemellata*. This species and *C. hyperici* were introduced from Europe to control Klamath weed, which was a serious rangeland pest in the Pacific Northwest. *Chrysolina quadrigemina* succeeded in controlling this weed and is now the more common of the two beetles.

Chrysomela aeneicollis

ADULT Wingcovers highly variable, ranging from black with red or pale brown spots to pale brown with black spots (spots usually connected); pronotum black. **BODY LENGTH** 6 mm. **LARVA** Shiny black with white glands along sides. **FOOD** Adult and larva: Coastal willow (*Salix hookeriana*), narrow-leaved willow (*S. exigua*), and other willows (*Salix* spp.) (leaves). **FOUND** Very common throughout the region.

Warmer inland areas seem to have adults that are more brightly colored (shiny black with red or pale brown spots); cool coastal areas seem to have paler individuals. Adults with different wingcover patterns may occur together. Skeletonized leaves are evidence of larval damage. When disturbed, the larva exudes a liquid from the glands along its sides.

Chrysomela confluens

COTTONWOOD LEAF BEETLE

ADULT Wingcovers with two distinct color patterns—iridescent purple-brown to dark blue, or gold with iridescent purple-brown to dark blue spots; pronotum black with red margins, each with one black spot. **BODY LENGTH** 8 mm. **LARVA** Shiny black with white glands along sides. **FOOD** Adult and larva: Willow (*Salix* spp.), poplar (*Populus* spp.). **FOUND** Very common throughout the region.

Adults with either wingcover color pattern often occur together. When disturbed, the larva exudes a liquid from the glands along its sides.

Chrysomela mainensis

ALDER LEAF BEETLE

ADULT Wingcovers orangish red, each with seven black spots that are sometimes coalesced; pronotum black with orangish brown lateral margins, each with one tiny black spot. **BODY LENGTH** 7 mm. **LARVA** Shiny black with white glands along sides. **FOOD** Adult and larva: Red alder (*Alnus rubra*) and other alders (*Alnus* spp.). **FOUND** Throughout the region.

Also known as *Chrysomela interna*.

Chrysomela aeneicollis, larvae

Chrysomela confluens, mating pair (female laying eggs)

Chrysomela confluens

Chrysomela mainensis, adult and larvae

Chrysomela schaefferi

Cryptocephalus castaneus

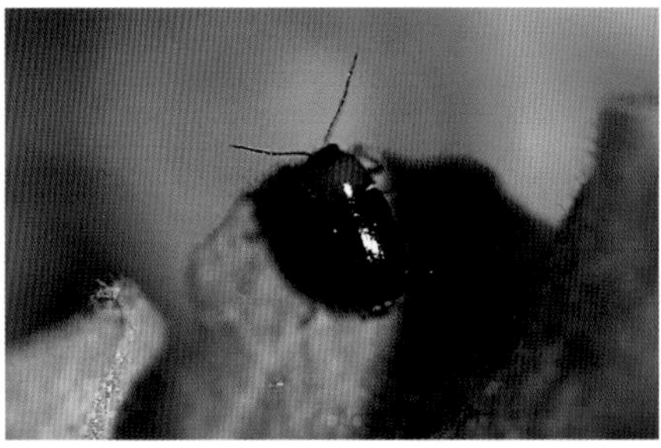

Cryptocephalus sanguinicollis

Diabrotica undecimpunctata

Chrysomela schaefferi

SCHAEFFER'S LEAF BEETLE

ADULT Wingcovers reddish brown; pronotum black with white lateral margins, each with one tiny black spot. **BODY LENGTH** 6 mm. **FOOD** Adult and larva: Willow (*Salix* spp.). **FOUND** Throughout the region.

This is the least common of the four *Chrysomela* species found in the region.

Cryptocephalus castaneus

CASTANEOUS LEAF BEETLE

ADULT Wingcovers white with reddish brown blotches of varying sizes; pronotum white with reddish brown longitudinal lines. **BODY LENGTH** 6 mm. **FOOD** Adult: Narrow-leaved willow (*Salix exigua*) and other willows (*Salix* spp.). **FOUND** Range unknown, but Pete has collected this beetle in Oregon and California.

Cryptocephalus sanguinicollis

ADULT Small but very distinctive; wingcovers and head black; pronotum red. **BODY LENGTH** 5 mm. **FOOD** Adult: California-lilac (*Ceanothus* spp.), willow (*Salix* spp.), wild buckwheats (*Eriogonum compositum*, *E. nudum*, and sulfur flower [*E. umbellatum*]), and other plants. **FOUND** Throughout the region. May be very common on wild buckwheat and California-lilac flowers in the spring.

Diabrotica undecimpunctata

WESTERN SPOTTED CUCUMBER BEETLE

ADULT Wingcovers shiny yellow to yellowish green, each with six black spots. **BODY LENGTH** 6 mm. **FOOD** Adult: Plants (foliage and flowers). Larva: Plants (roots). **FOUND** From Washington to California. In mild coastal areas adults may be found any month of the year feeding on light-colored flowers (e.g., dandelions). Common in vegetable gardens spring through summer.

Disonycha alternata

ADULT Wingcovers with white and black lines; pronotum red with four black spots; head and abdomen red; back legs enlarged. **BODY LENGTH** 8 mm. **FOOD** Adult: Narrow-leaved willow (*Salix exigua*) and other willows (*Salix* spp.). **FOUND** Throughout the region. Pete has found adults only on willows that are in close proximity to water.

Also known as *Disonycha quinquevittata*. This species uses its back legs for jumping like a flea when pursued by a predator or photographer.

Disonycha alternata

Galeruca rudis

LUPINE LEAF BEETLE

ADULT Wingcovers black, ridged, and with pale yellow margins. **BODY LENGTH** 11 mm. **LARVA** Dark brown to black. **FOOD** Adult and larva: Yellow bush lupine (*Lupinus arboreus*) and other lupines (*Lupinus* spp.) (foliage). **FOUND** Throughout the region, from coastal areas to mountain meadows.

Pete has seen large patches of lupine plants that have been totally defoliated by the adults and larvae.

Galeruca rudis

Gastrophysa cyanea

GREEN DOCK BEETLE

ADULT Wingcovers and pronotum iridescent green (may also appear iridescent blue or black, depending on angle of light). **BODY LENGTH** 5 mm. **FOOD** Adult and larva: Dock (*Rumex* spp.) (leaves). **FOUND** Throughout the region. In the spring look for clusters of orange eggs on dock leaves.

This species was introduced from Europe.

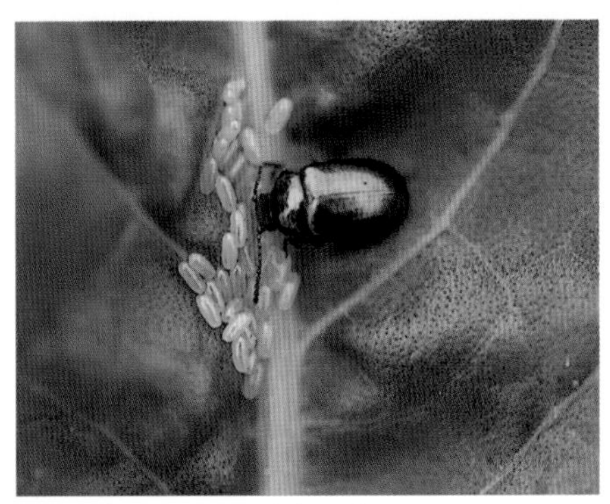
Gastrophysa cyanea, adult with eggs

Jonthonota nigripes

Jonthonota nigripes, larva

Metriona bicolor

Pachybrachis circumcinctus

Jonthonota nigripes

BLACK-LEGGED TORTOISE BEETLE

ADULT Wingcovers red with flat, translucent margins and each with three black spots; pronotum red with flat, translucent margins and completely covers head. **BODY LENGTH** 7 mm. **LARVA** Yellow, broadly flat, and covered with spines; carries its own excrement above its body like a parasol using an appendage aptly called the *anal fork*. Similar to *Metriona bicolor*. **FOOD** Adult and larva: Morning-glory (*Calystegia* spp.). **FOUND** California and Oregon and probably Washington (Hatch [1957–1971] lists this species but not its range).

This is a special insect for two reasons: the adult looks like a burnished jewel, and the larva looks fierce with its spiny body and its excrement "proudly" held aloft.

Metriona bicolor

GOLDEN TORTOISE BEETLE

ADULT Wingcovers iridescent gold to greenish gold with flat, translucent margins and each with three black spots; pronotum iridescent gold to greenish gold with flat, translucent margins and completely covers head. **BODY LENGTH** 6 mm. **LARVA** Yellow, broadly flat, and covered with spines; carries its own excrement above its body like a parasol using an appendage aptly called the *anal fork*. Similar to *Jonthonota nigripes*. **FOOD** Adult and larva: Morning-glory (*Calystegia* spp.). **FOUND** Throughout the U.S.

Pachybrachis circumcinctus

ADULT Wingcovers pale white, each with longitudinal black band and small black spot on shoulder; pronotum pale yellow with large brown spot. **BODY LENGTH** 5 mm. **FOOD** Adult: Sitka willow (*Salix sitchensis*), narrow-leaved willow (*S. exigua*). **FOUND** Throughout the region.

Saxinis saucia

RED-SHOULDERED LEAF BEETLE

ADULT Wingcovers dark blue to black, each with red spot on shoulder. **BODY LENGTH** 6 mm. **FOOD** Adult: *Eriogonum compositum, E. nudum,* and other wild buckwheats (*Eriogonum* spp.). Larva: Material in ant nests. **FOUND** Throughout the region. Adult quite common on flowers in the spring.

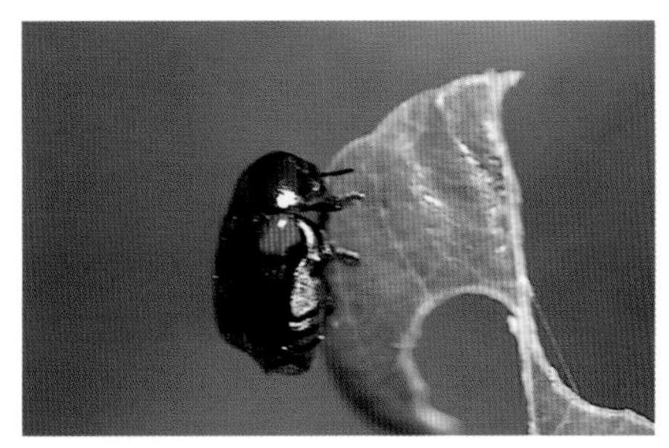

Saxinis saucia

Timarcha intricata

ADULT Wingcovers and pronotum black, shiny, and distinctly pitted; flightless. **BODY LENGTH** 10 mm. **FOOD** Adult: Beach strawberry (*Fragaria chiloensis*) and probably other plants as well. **FOUND** Throughout the region.

Timarcha intricata

Tricholochmaea punctipennis

WILLOW LEAF BEETLE

ADULT Wingcovers black; pronotum yellow or orange with three black spots. **BODY LENGTH** 6 mm. **LARVA** Black, slender. **FOOD** Adult and larva: Willow (*Salix* spp.), alder (*Alnus* spp.). **FOUND** Throughout the region, often in high numbers.

Now known as *Pyrrhalta punctipennis.*

Tricholochmaea punctipennis

Tricholochmaea punctipennis, larvae

Trirhabda sp.

Trirhabda spp.

ADULT Wingcovers iridescent green; pronotum yellow with three black spots. **BODY LENGTH** 9 mm. **LARVA** Iridescent green. **FOOD** Adult: Coast goldenrod (*Solidago spathulata*) and other goldenrods (*Solidago* spp.). **FOUND** Throughout the region.

Although this genus is common in the region, it is very difficult to identify to species.

■ TIGER BEETLES Family Cicindelidae

The tiger beetles are active insects that have long legs and are commonly brightly colored, usually iridescent, and distinctly patterned. The patterns are an important characteristic in identifying these beetles to species. They have medium- to large-sized, elongate, cylindrical bodies and large, sickle-shaped jaws. They are usually found in sunny areas, often on sandy beaches or in areas of dry soil. They can run and fly rapidly and are very wary and difficult to approach. When handled, they can give a painful bite. Both the adults and larvae aggressively prey on a variety of small invertebrates. The larvae live in vertical underground burrows dug in sandy or dry soil; they wait at the entrance of their respective burrow to capture passing prey. When an invertebrate passes by a burrow entrance, the larva springs out, grabs the prey, and drags it back into the burrow.

Cicindela oregona

Cicindela oregona

OREGON TIGER BEETLE

ADULT Wingcovers dark brown with green iridescence and white spots and bands. **BODY LENGTH** 13 mm. **LARVA** Pale white; head flattened. **FOOD** Adult and larva: Invertebrates.

Cicindela oregona, larva

FOUND Throughout the region, most commonly in sandy areas.

The adults congregate in large numbers. They search for prey by flying very low over the ground.

Cicindela purpurea

Cicindela purpurea

ADULT Green; wingcovers each with yellow medial line, small yellow spot, and yellow tip. **BODY LENGTH** 15 mm. **FOOD** Adult and larva: Invertebrates. **FOUND** Throughout the region.

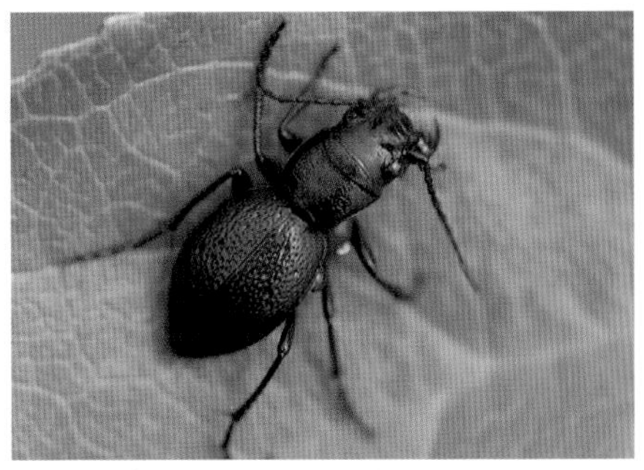
Omus audouini

Omus audouini

FLIGHTLESS TIGER BEETLE

ADULT Black; pronotum with sculpted margins; wingcovers fused, with a few small pits interspersed among many tiny pits; hindwings absent. **BODY LENGTH** 20 mm. **FOOD** Adult and larva: Invertebrates. **FOUND** Throughout the region. Adults hide under objects during the day; most active at night.

■ CHECKERED BEETLES Family Cleridae

The adults of most species in this family are characterized by a brightly colored and/or distinctly patterned (hence the common name), densely hairy body and large head. In all species, the adult body is small and elongate. The adults and larvae of most species are predaceous and are commonly found under bark. The larvae of a few species feed on grasshopper eggs or bee larvae. Forest managers consider these beetles to be beneficial because they prey on bark beetles and on the larvae of other wood-boring insects. There are some species, however, that are regarded as pests because they feed on stored food products.

Chariessa elegans

ADULT Head, pronotum, abdomen, and legs dark red; wingcovers black to bluish black; antennae with last three segments clubbed. **BODY LENGTH** 13 mm. **FOOD** Adult and larva: Invertebrates. **FOUND** Throughout the region. Larvae found under tree bark.

Chariessa elegans

Enoclerus eximius

Enoclerus eximius

ADULT Wingcovers red, each with one shared black spot at base, one large, black spot at tip, and white pubescent band near tip. **BODY LENGTH** 10 mm. **FOOD** Adult and larva: Invertebrates. **FOUND** Throughout the region, often on plants along waterways.

Enoclerus sphegeus

Enoclerus sphegeus

WASPLIKE CLERID

ADULT Black, hairy; wingcovers each with broad, grayish white band across middle; abdomen red. **BODY LENGTH** 12 mm. **FOOD** Adult and larva: Invertebrates. **FOUND** Throughout the region.

This species is considered a significant predator on bark beetles in western U.S. coniferous forests. Attempts to raise large numbers in the laboratory and release them for biological control have failed.

Trichodes bimaculatus

Trichodes bimaculatus

TWO-SPOTTED CLERID

ADULT Wingcovers blue-black, each with red spot. Similar to *Trichodes oregonensis* (not described in this field guide) except wingcovers of latter with no red spots, or, if present, very small. **BODY LENGTH** 10 mm. **FOOD** Adult: Invertebrates. Larva: Wasp and bee larvae and their stored food. **FOUND** Oregon and California. Look for adult on flowers.

Trichodes ornatus, typical pattern

Trichodes ornatus

ORNATE CHECKERED BEETLE

ADULT Wingcovers blue-black and variably patterned with yellow patches and/or bands. **BODY LENGTH** 13 mm. **FOOD** Adult: Invertebrates. Larva: Immature stages of bees and wasps. **FOUND** Throughout the region. Look for adult on or inside the flower clusters of such plants as wild buckwheat (*Eriogonum* spp.) or California-lilac (*Ceanothus* spp.). Larvae live inside bee and wasp nests.

This is one of the most common flower-inhabiting beetles in the spring and early summer.

■ LADYBIRD BEETLES Family Coccinellidae

This family of beetles, which are often called *ladybugs* in the United States, is large and diverse; there are 57 genera and 475 species in the United States and Canada. They are among the most familiar of beetles, easily recognized by their round, convex, often spotted wingcovers. Most are shiny red, orange, brown, or black with various markings. The head is hidden from above by the expanded pronotum. Some species have striking color patterns but are difficult to identify because these patterns greatly vary within any given population; in fact, the differences in dorsal color and pattern can be greater within a species than between species. (The only way to accurately identify polymorphic species is by observing them in the field over a period of years or by examining a museum's pinned collection that contains a comprehensive representation of individuals within the species.) The adults are common on plants and frequently overwinter in large hibernating aggregations. The adults and larvae of most species are predaceous, mostly on other small invertebrates; the adults are typically common in the spring, mating and feeding on large populations of aphids. The members of one genus feed on fungi. The larvae are elongate, rather flattened, and usually covered with tiny tubercles or spines and brightly colored spots and bands. They are usually found in or near aphid colonies. In warm climates, there are many generations per year. Ladybird beetles are considered to be beneficial; they are commonly used as a biological control of insect pests (many species have been imported in an attempt to control crop pests).

Human societies have long regarded ladybird beetles in a favorable light. Written accounts of these beetles go back to at least fifteenth-century Europe, where the beetles were (and still are) often associated with the Virgin Mary. During the Middle Ages, ladybird beetles were instrumental in controlling grapevine pests and consequently were given their common name in honor of "Our Lady." In Germany they are referred to as *Lady-beetles of the Virgin Mary* and in France as *Animals of the Virgin*.

Ladybird beetles were also thought to have supernatural powers that ranged from curing illnesses such as measles and colic or a toothache (when mashed and applied to the tooth) to helping the lovelorn. These supernatural powers were often celebrated in rhymes; in the following example, a ladybird beetle is called on to help the woman find a husband and indicate from which direction her betrothed will arrive:

> This Ladyfly I take from off the grass
> Whose spotted back might scarlet red surpass.
> Fly, Ladybird, north, south, or east or west,
> Fly where the man is found that I love best.
> He leaves my hand, see to the west he's flowen,
> To call my true-love from the faithless town.

It was not until the early 1900s that ladybird beetles were officially considered for use as a biological control on insect pests in the United States. The U.S. Bureau of Entomology, after evaluating the release of the beetles into agricultural fields, decided that ladybird beetles were not an effective method of insect control. Even with the release of tens of thousands of beetles per acre, many of the beetles rapidly dispersed into areas other than the release areas, so that within the release areas, the ladybird beetle populations, if any, remained the same or increased only slightly.

In California, in the early 1900s, large numbers of convergent ladybird beetles (*Hippodamia convergens*) were collected from hibernating aggregations and shipped by train to aphid-infested agricultural fields. In 1914 approximately two tons of beetles were shipped to growers in the Imperial Valley. At that time the railroad companies did not impose shipping charges on the growers. Despite the fact that 50 to 75 percent of the beetles died during shipping, growers were very eager to order more shipments of beetles until the railroads began to ask the growers to pay freight charges, albeit minimal. The growers then decided that ladybird beetles were not worth the charges and stopped ordering shipments.

Ladybird beetles have been popular, nevertheless, in controlling insects, such as aphids, especially among gardeners. Their popularity has increased as the public has become increasingly aware of the negative impacts of insecticides. There is an unfortunate side effect to this popularity, however: the commercial exploitation of ladybird beetles. The uncontrolled collecting of hibernating (overwintering) ladybird beetles can be devastating to native ladybird beetle populations. And those innocent-looking mesh bags full of ladybird beetles one sees hanging from the counter stand in the local garden shop actually represent a cruel and unconscionable practice: ladybird beetles sold in retail stores are usually exposed to high temperatures, low humidity, and no food for weeks. Even if they survive until bought and released, they are often so weakened, they die soon after being released.

The best way to have ladybird beetles in one's garden is to attract resident ladybird beetles by providing a ladybird beetle–friendly environment for them. One can do this by planting native plants that will attract ladybird beetles and by not using insecticides.

Adalia bipunctata, typical pattern

Adalia bipunctata

TWO-SPOTTED LADYBIRD BEETLE

ADULT Coloration and pattern extremely variable; most commonly, wingcovers red, each with one black spot, and pronotum black with white patch on each side. **BODY LENGTH** 5 mm. **FOOD** Adult and larva: Invertebrates. **FOUND** Common in most of the U.S. Look for adults and larvae wherever large infestations of aphids occur, such as on European birch (*Betula pendula*), which are very susceptible to heavy infestations of birch aphids.

Adalia bipunctata

Adalia bipunctata

Because this species is so variable and widespread, some authors have broken it down into several species. According to Gordon (1985), however, *Adalia* has only one species, *A. bipunctata*, in North America.

Anatis rathvoni

RATHVON'S LADYBIRD BEETLE

ADULT Wingcovers brownish red with black spots; pronotum white with large, black central area within which are two white spots. **BODY LENGTH** 10 mm. **LARVA** Black and orange and covered with spines. **FOOD** Adult and larva: Invertebrates. **FOUND** Common throughout the region. In some years, high numbers may be found feeding on conifer and oak aphids in spring and early summer.

This is one of the largest ladybird beetles in the region.

Anatis rathvoni

Brachiacantha blaisdelli

BLAISDEL'S LADYBIRD BEETLE

ADULT Wingcovers black, each with orange medial band and orange spot near tip; pronotum black with orange anterior and lateral margins. **BODY LENGTH** 5 mm. **FOOD** Adult and larva: Invertebrates. **FOUND** Oregon, California, and Nevada.

Brachiacantha blaisdelli

Calvia quatuordecimguttata

Calvia quatuordecimguttata

Chilocorus orbus

Calvia quatuordecimguttata

FOURTEEN-SPOTTED LADYBIRD BEETLE

ADULT Coloration and pattern extremely variable; two most common variations—wingcovers black, each with large, reddish pink spot, and pronotum black; and wingcovers reddish pink and covered with black spots and pronotum reddish pink with two large black spots. **BODY LENGTH** 5 mm. **FOOD** Adult and larva: Invertebrates. **FOUND** From Alaska to San Francisco, California. Look for adults in early spring on willows with heavy infestations of psyllids (insects that look and feed like aphids), a favorite food of this species. Adults can normally be found on the same trees every spring.

This species is one of the best examples of color and pattern variability within a species; coleopterists have assigned it 48 synonyms. Pete saw, on one tree, a group of adults comprising at least four different dorsal color patterns, with individuals apparently selecting breeding mates regardless of their color pattern.

Chilocorus orbus

TWO-STABBED LADYBIRD BEETLE

ADULT Oval; wingcovers black, each with red spot. **BODY LENGTH** 5 mm. **FOOD** Adult and larva: Invertebrates, preferably scale insects. **FOUND** From Washington to southern California.

Pete found this species in large numbers in the fall, feeding on aphids on Oregon oak (*Quercus garryana*) along the Columbia River in Washington and central and eastern Oregon.

Coccinella californica

CALIFORNIA LADYBIRD BEETLE

ADULT Wingcovers red with suture edge between wingcovers dark brown; pronotum black with white patch on each side. **BODY LENGTH** 6 mm. **FOOD** Adult and larva: Invertebrates. **FOUND** Common along the coast from British Columbia to southern California.

Coccinella californica

Coccinella monticola

ADULT Wingcovers orange to pale orange with black spots; pronotum black with pale orange patch on each side; head black with two pale orange spots. **BODY LENGTH** 7 mm. **FOOD** Adult and larva: Invertebrates. **FOUND** Throughout the region. Common, except in California.

Coccinella monticola

Coccinella septempunctata

SEVEN-SPOTTED LADYBIRD BEETLE

ADULT Wingcovers orangish red, each with three black spots and one shared black spot at base. **BODY LENGTH** 7 mm. **FOOD** Adult and larva: Invertebrates. **FOUND** Common throughout the region.

Very little color or pattern variation in the adults of this species makes it one of our easiest ladybirds to identify. It was introduced from Europe for biological control of aphids. This species and another introduced species, the Asian spotted ladybird beetle (*Harmonia axyridis*), have come to dominate some urban populations of ladybird beetles.

Coccinella septempunctata

Coccinella transversoguttata

ADULT Wingcovers brownish orange to pale orange with black spots; pronotum black with white spot on each side. **BODY LENGTH** 7 mm. **FOOD** Adult and larva: Invertebrates. **FOUND** Throughout the region.

Coccinella transversoguttata

Coccinella trifasciata, typical pattern

Coccinella trifasciata

THREE-BANDED LADYBIRD BEETLE

ADULT Wingcover coloration and pattern extremely variable—red to pale orange, most commonly each with three black transverse bands, or with black bands reduced to spots; pronotum black with white anterior and lateral margins. **BODY LENGTH** 5 mm. **FOOD** Adult and larva: Invertebrates. **FOUND** Throughout much of the U.S.

Coccinella trifasciata

Cycloneda polita

WESTERN BLOOD-RED LADYBIRD BEETLE

ADULT Wingcovers usually deep red; pronotum black with one incomplete, white ring on each side. **BODY LENGTH** 6 mm (usually less). **FOOD** Adult and larva: Invertebrates. **FOUND** From Canada to southern California.

Cycloneda polita

Exochomus californicus

ADULT Wingcovers black, each with orange shoulder and orange spot near tip; pronotum black. **BODY LENGTH** 4 mm. **FOOD** Adult and larva: Invertebrates. **FOUND** From southwestern Washington to California.

Harmonia axyridis

ASIAN SPOTTED LADYBIRD BEETLE

ADULT Wingcovers red to orange with black spots; pronotum white with black M in middle. Newly emerged adult is paler and has no

Exochomus californicus

Harmonia axyridis, newly emerged adult and mature adult

spots. **BODY LENGTH** 7 mm. **LARVA** Dark brown and spiny; abdomen with row of orange spots on each side. **FOOD** Adult and larva: Invertebrates. **FOUND** Throughout the region. Often enters buildings in large numbers to overwinter.

Since its introduction from Japan circa 1978 to control aphids, it has been expanding its range throughout the U.S. to become one of our most common ladybirds.

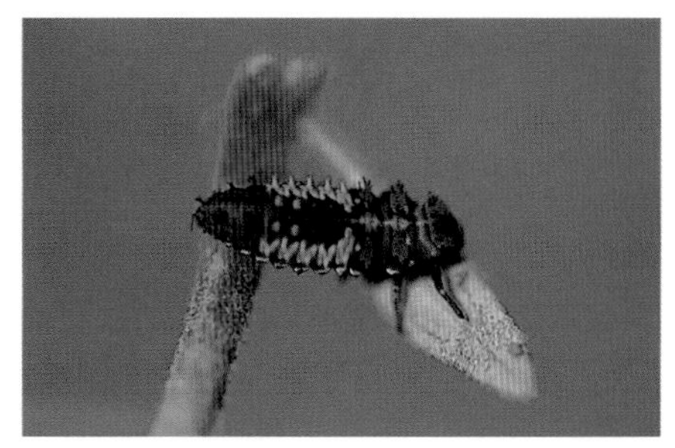

Harmonia axyridis, larva

Hippodamia apicalis

ADULT Wingcovers red to orange, each with three black spots and one shared black basal spot, spots at tip touch suture and may coalesce; pronotum black with white posterior spot and white anterior and lateral margins. **BODY LENGTH** 4 mm. **FOOD** Adult and larva: Invertebrates. **FOUND** Throughout the region.

Hippodamia apicalis

Hippodamia convergens

CONVERGENT LADYBIRD BEETLE

ADULT Wingcover color and pattern variable but usually red to orange with black spots, often six per wingcover; pronotum black with pair of white lines and white anterior and lateral margins. Pronotum is the best feature to use to identify this species since it is more consistent than wingcover pattern. **BODY LENGTH** 7 mm. **FOOD** Adult and larva: Invertebrates. **FOUND** Throughout the U.S.

This species is the most common large ladybird beetle in the region.

Hippodamia convergens

Hippodamia convergens, overwintering adults

Hippodamia lunatomaculata

Hippodamia moesta

Hippodamia parenthesis

Hippodamia sinuata

Hippodamia lunatomaculata

ADULT Wingcovers red to orange, each with three black spots and one shared black basal spot; pronotum black with white posterior spot and white anterior and lateral margins. **BODY LENGTH** 5 mm. **FOOD** Adult and larva: Invertebrates. **FOUND** Oregon and California.

Hippodamia moesta

ADULT Wingcovers black, each with one red to orange spot near tip; pronotum black with pair of white spots and white anterior and lateral margins. **BODY LENGTH** 7 mm. **FOOD** Adult and larva: Invertebrates. **FOUND** From British Columbia to northern California.

Hippodamia parenthesis

PARENTHESIS LADYBIRD BEETLE

ADULT Wingcovers red to orange, each with three black spots and one shared black basal spot, spots at tip usually coalesced into parenthesis shape; pronotum black with white posterior spot and white anterior and lateral margins. **BODY LENGTH** 5 mm. **FOOD** Adult and larva: Invertebrates. **FOUND** Throughout most of the U.S.; on West Coast occurs in eastern Washington, eastern Oregon, and California (except in northernmost part of state).

Hippodamia sinuata

ADULT Wingcovers red to orange, each with three black spots and one shared black basal spot; pronotum black with pair of white lines and white anterior and lateral margins. **BODY LENGTH** 5 mm. **FOOD** Adult and larva: Invertebrates. **FOUND** From Alaska to northern California.

Mulsantina picta

PINE LADYBIRD BEETLE

ADULT Wingcovers whitish pale brown with variable dark brown markings; pronotum white with black M-shaped mark. **BODY LENGTH** 5 mm. **FOOD** Adult and larva: Invertebrates. **FOUND** Throughout the U.S.

Mulsantina picta

Olla v-nigrum

ASHY GRAY LADYBIRD BEETLE

ADULT Wingcovers with two distinct color patterns—pale yellow with small, black spots, or black, each with one large, red spot. **BODY LENGTH** 6 mm. **FOOD** Adult and larva: Invertebrates. **FOUND** From British Columbia to California; most common in California.

The two completely different wingcover patterns have generated many synonyms of this species in the literature (e.g., Gordon [1985] cites 22 synonyms).

Olla v-nigrum

Olla v-nigrum

Psyllobora borealis

SMALL ASHY GRAY LADYBIRD BEETLE

ADULT Wingcovers creamy white, each with separate and coalesced black spots, discrete black spot mid-wingcover on outer margin; pronotum creamy white with black spots. Very similar to *Psyllobora vigintimaculata* except wingcovers of latter each without discrete black spot mid-wingcover on outer margin. **BODY LENGTH** 3 mm. **FOOD** Adult and larva: Fungus. **FOUND** Throughout the U.S. May occur in very high numbers from late summer through the fall on willow leaves that are infected with mildew.

This species and *Psyllobora vigintimaculata* are probably our most common ladybird beetles and are the only ladybird beetles in the region that are not predaceous.

Psyllobora vigintimaculata and *P. borealis*

Psyllobora vigintimaculata

TWENTY-SPOTTED LADYBIRD BEETLE

ADULT Wingcovers creamy white with separate and coalesced black spots; pronotum creamy white with black spots (see p. 65). Very similar to *Psyllobora borealis* except wingcovers of latter each with discrete black spot mid-wingcover on outer margin. **BODY LENGTH** 3 mm. **FOOD** Adult and larva: Powdery mildew (*Sphaerotheca* spp., *Podosphaera* spp.). **FOUND** Throughout the U.S. May occur in very high numbers from late summer through the fall on willow leaves that are infected with mildew.

This species and *Psyllobora borealis* are probably our most common ladybird beetles and are the only ladybird beetles in the region that are not predaceous.

Scymnus spp.

ADULT Tiny; black. **BODY LENGTH** 2 mm. **LARVA** Looks like a mealybug—dark gray and covered with white, waxy threads. **FOOD** Adult and larva: Invertebrates. **FOUND** Throughout the region. To find the larvae, look for heavy infestations of aphids not attended by ants; the larvae can be observed eating the aphids.

Scymnus sp., larvae with aphids

This genus includes many species, most of which have adults that are small and black, making them a very difficult group for amateurs to identify.

■ SNOUT BEETLES Family Curculionidae

Also known as *weevils,* these beetles make up the largest family of insects, with over 40,000 known species worldwide and over 2600 known in North America. The hard-bodied adults vary in size and shape but all are distinguished by a head that is elongated into a slender, downcurved snout, with elbowed or clubbed antennae attached partway down and jaws at the tip. These beetles are found almost everywhere. All but a few species are plant feeders. The larvae are C-shaped, and most live within the tissues of their host plants. Both the adults and larvae of some species are major agricultural and garden pests.

Otiorhynchus sulcatus

Otiorhynchus sulcatus

BLACK VINE WEEVIL

ADULT Wingcovers fused, dark brown to black with small, scattered yellow to golden brown spots and covered with parallel rows of tiny pits; hindwings absent. **BODY LENGTH** 11 mm. **LARVA** Shiny white. **FOOD** Adult: Plants (foliage). Larva: Plants (underground roots). **FOUND** Throughout the U.S.

In the U.S. the females of this introduced species reproduce parthenogenetically.

Otiorhynchus sulcatus, larvae

Rhynchites bicolor

ROSE WEEVIL

ADULT Wingcovers and pronotum dark red; head and snout black. **BODY LENGTH** 6 mm. **FOOD** Adult and larva: Preferably thimbleberry (*Rubus parviflorus*) but also other berries (*Rubus* spp.) and *Rosa* spp. **FOUND** Throughout the U.S.

Also known as *Merhychites bicolor*.

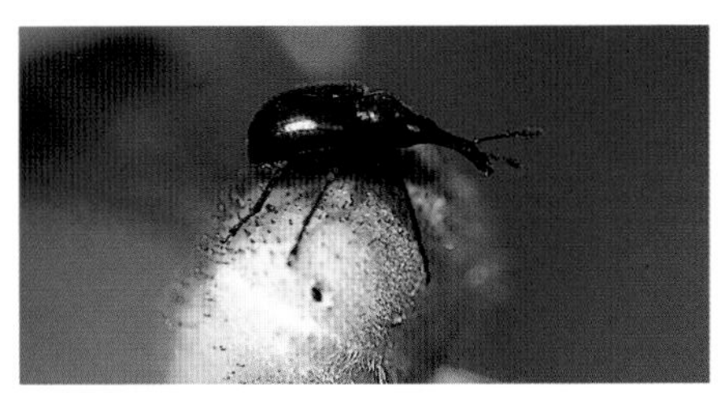

Rhynchites bicolor

■ PREDACEOUS DIVING BEETLES Family Dytiscidae

This large family of aquatic beetles is most common in ponds and slow-moving streams. The adults have smooth, oval, hard bodies that are usually black or brown in color. The hind legs are flattened and fringed with long hairs, which help propel them through the water. The adults obtain air from the surface of the water, often hanging at the surface with their head downward. A chamber under their forewings where air can be stored allows them to stay submerged for long periods. The adults may leave the water at night and fly around; many species are attracted to lights at night. Dytiscids are long-lived, often hibernating in bottom mud over the winter. Both the adults and larvae are predaceous, feeding on a variety of aquatic animals, including fish. The larvae, often called *water tigers*, have long, sicklelike jaws and can attack prey larger than themselves.

Acilius semisulcatus

SMALL FLAT DIVING BEETLE

ADULT Wingcovers dark brown, each with four longitudinal grooves; pronotum with two dark brown transverse lines; head with interrupted M between eyes. **BODY LENGTH** 14 mm. **FOOD** Adult and larva: Invertebrates. **FOUND** Throughout the region.

Acilius semisulcatus

CLICK BEETLES Family Elateridae

Although this large family has many species that commonly occur in the Pacific Northwest, it is not well documented in the literature and not well represented in collections. These beetles are noted for their ability to escape by flipping themselves into the air and at the same time making a sharp clicking sound. The adults are elongate with the posterior corners of the pronotum lengthened into points or spines. In most species, the adults are medium-sized and black or brown in color (some are brightly colored). They are usually found on flowers or bark. The adults of some species are plant feeders; others are predaceous. The larvae are commonly called *wireworms* because of their slender, hard, shiny bodies. In most species, the larvae live in the soil or decaying wood; they feed on the roots of plants, and some are agricultural and garden pests.

Alaus melanops

Ampedus apicatus

Alaus melanops

EYED ELATER

ADULT Black with white speckling; pronotum with two velvety black eyespots. **BODY LENGTH** 40 mm. **FOOD** Adult and larva: Invertebrates. **FOUND** Throughout the region.

Ampedus apicatus

ADULT Pronotum black; wingcovers orangish red, each with black apical patch reaching margin only at tip. **BODY LENGTH** 12 mm. **FOOD** Adult and larva: Unknown. **FOUND** Throughout the region. Adults usually under tree bark or occasionally on flowers.

The *Ampedus* spp. constitute a large genus that is well represented in the region; however, there are few specimens in collections, and this genus is not well documented in the literature.

Ctenicera silvaticus

ADULT Black; pronotum finely pitted and with pointed marginal tips; wingcovers with longitudinal grooves. **BODY LENGTH** 18 mm. **LARVA** Shiny orange. **FOOD** Adult: Probably plants. Larva: Plants (probably grasses) (roots). **FOUND** Throughout the region. Early to midsummer, adults can be seen in mating flights several inches above grasses in pastures and mountain meadows. (Birds often pursue the hovering adult beetles.)

The *Ctenicera* spp. constitute a large genus that is well represented in the region; however, there are few specimens in collections, and this genus is not well documented in the literature.

Ctenicera silvaticus

Lacon profusa

ADULT Resembles tree bark—dark brown with light brown speckling, some of which is coalesced. **BODY LENGTH** 12 mm. **FOOD** Adult and larva: Unknown. **FOUND** Throughout the region.

Lacon profusa

■ HANDSOME FUNGUS BEETLES Family Endomychidae

Little is known about these smooth, shiny, brightly colored beetles. They are similar to the ladybird beetles (Coccinellidae) in appearance but have a head that is easily visible from above and a pronotum with two longitudinal grooves. Most species are found under bark in rotting wood, in fungi, and in decaying fruit; they feed on fungi.

Mycetina idahoensis

IDAHO HANDSOME FUNGUS BEETLE

ADULT Pronotum black with red lateral margins; wingcovers black, each with two red spots. **BODY LENGTH** 4 mm. **FOOD** Adult: Fungi. **FOUND** Throughout the region. Look for adults in spring around downed trees.

Mycetina idahoensis

MARSH BEETLES Family Helodidae (Scirtidae)

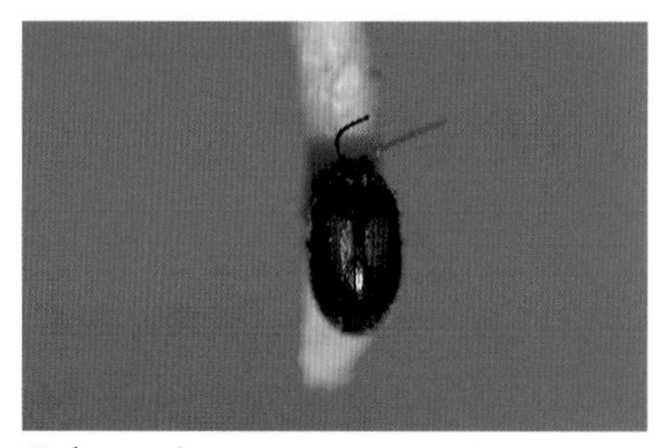

Cyphon concinna

The members of this family are tiny, oval beetles that occur on vegetation in wetlands. The larvae are aquatic.

Cyphon concinna

ADULT Wingcovers black with red base. **BODY LENGTH** 4 mm. **FOOD** Adult and larva: Unknown. **FOUND** Throughout the region.

CLOWN BEETLES Family Histeridae

Clown beetles, also known as *hister beetles,* are small, hard, compact, broadly oval beetles that are usually shiny black. The antennae are elbowed and clubbed. In most species, the wingcovers are shortened, exposing the tip of the abdomen, and have a squared posterior margin. When disturbed, the adult can retract its appendages so closely to its body that it is often difficult for other insects, like ants, to attack it. Most species are found in or near decaying organic matter; others live in fungus, under bark, or in ant nests. Both the adults and larvae of most species prey on other small invertebrates. The larvae of most species are found in the tunnels of wood-boring beetles and feed on their larvae.

Spilodiscus sellatus

Spilodiscus sellatus

ADULT Shiny black; wingcovers each with wide, L-shaped, red area. **BODY LENGTH** 13 mm. **FOOD** Adult: Probably ants. Larva: Possibly ant larvae. **FOUND** Throughout the region. Adult found in sandy areas, larva in ant nests.

■ FIREFLY BEETLES Family Lampyridae

Also known as *lightning bugs*, many members of this well-known family have light organs at the tip of the abdomen. Although few species occur in the Pacific Northwest, and none of the winged adults of these species glow, those that do occur here are common. The winged adults in this family have an elongated, soft body with the head covered by the pronotum. The females of many species are wingless and larviform. The adults are usually found on vegetation during the day. They feed little or not at all; some are predaceous. The larvae prey on smaller insects and snails. In many species, they are luminescent and, along with the luminescent larviform females of some species, are often called *glowworms*.

Ellychnia hatchi

BLACK LAMPYRID

ADULT Black; pronotum with flattened, red lateral margins. **BODY LENGTH** 16 mm. **FOOD** Adult: Probably none. Larva: Invertebrates. **FOUND** From British Columbia to California, in wet areas, commonly in spring and early summer.

There are a number of species in this genus, and one needs to use the identification key in Fender (1970) to identify them. Neither the adults nor larvae glow.

Ellychnia hatchi

Matheteus theveneti

ADULT Wingcovers red; pronotum red with large, central, black spot and flattened margins; body and legs black; antennae black, each segment with narrow, toothlike projection. **BODY LENGTH** 11 mm. **FOOD** Adult and larva: Invertebrates. **FOUND** Oregon and California.

Some entomologists place this species in the family Omethidae.

Matheteus theveneti

Microphotus angustus

PINK GLOWWORM

ADULT Male, wingcovers and pronotum grayish brown; thorax and abdomen pink. Female (and larva), larviform, flattened; creamy white tinged with pink; without wings or wingcovers; glows. **BODY LENGTH** 15 mm.

Microphotus angustus, female

FOOD Adult: Probably none. Larva: Invertebrates. **FOUND** Oregon and California. Adult in dry grassy areas in the summer. Adult female and larva hide under objects during the day.

Pterotus obscuripennis

ADULT Male, wingcovers black; pronotum reddish orange; head black and with long, branched antennae. Female (and larva), abdomen creamy white, black dorsally, and with two large, orangish yellow spots that glow near posterior end; without wings or wingcovers. Adult male similar to *Zarhipis integripennis*; latter's antennae more finely branched. **BODY LENGTH** Male 12 mm, female 35 mm. **FOOD** Adult: Probably none. Larva: Invertebrates. **FOUND** Throughout the region.

A resting male during the spring season can often be found slowly fanning the air with its antennae, trying to detect pheromones released by the females. The females do not use light to attract mates as many species of fireflies do. Their glow is probably used as a means of protection from predation: they are known to be distasteful to vertebrate predators, and their glow may warn potential predators to leave them alone much as the monarch butterfly does with its distinctive coloration and pattern. The glowworms, that is, the females and the larvae, are active at night, so to find them one must look at night (it would also be hard to see their "glow" in the light of day). They are most easily found on a warm spring or fall evening after or during a light rain. The best places to look for them are areas at ground level that are in complete darkness. The glowworms will appear as tiny "lights" on the ground. Stamping the ground with one's feet will sometimes make them start glowing. For more discussion of this species, see Dean (1979).

Pterotus obscuripennis, male

Pterotus obscuripennis, larva eating snail

STAG BEETLES Family Lucanidae

Stag beetles are so named because the males have large jaws that are sometimes branched like antlers. The females have smaller jaws. These beetles are sometimes called *pinching bugs* because the adults may pinch with their jaws. The adults are medium in size, and most are black or reddish brown. They have clubbed antennae, but the antennal segments that are enlarged do not close to form a compact club, as they do in the lamellicorn beetles (Scarabaeidae). Although there are probably fewer than 15 species in the Pacific Northwest, the adults are common in the spring and early summer. They are usually found on the ground in woods or on sandy beaches. They feed on leaves or honeydew or not at all. The white, C-shaped larvae live in decaying wood.

Ceruchus striatus

ADULT Shiny black; wingcovers with longitudinal grooves; head as wide as pronotum; antennae elbowed. **BODY LENGTH** 20 mm. **FOOD** Adult: Leaves, honeydew, or none. Larva: Rotting wood. **FOUND** Throughout the region. Look for adults in summer on tree trunks or rotting tree stumps.

Ceruchus striatus

Platyceroides agassii

AGASSIZ'S FLAT-HORNED STAG BEETLE

ADULT Reddish brown to black; wingcovers with fine, pitted longitudinal grooves; antennae elbowed. **BODY LENGTH** 10 mm. **FOOD** Adult: Leaves, honeydew, or none. Larva: Rotting wood. **FOUND** Oregon and California. Look for adults in early spring where there are large, downed trees.

Similar species occur throughout the region.

Platyceroides agassii

Sinodendron rugosum

RUGOSE STAG BEETLE

ADULT Male, black; head with short, distinctively curved (rhinoceroslike) horn. Female, similar to male but horn absent or, if present, smaller. **BODY LENGTH** 18 mm. **FOOD** Larva: Wood (probably decaying). **FOUND** Throughout the region.

Sinodendron rugosum, male

NET-WINGED BEETLES Family Lycidae

These medium-sized, elongate, soft-winged beetles somewhat resemble fireflies but can be distinguished from them by the network of raised lines on the wingcovers. The antennae are large with broad, flat segments. Most species are brightly colored with red, black, or yellow. The adults occur on foliage and tree trunks, usually in forested areas. Although they are known to feed on the juices of decaying plant material and occasionally on other insects, they probably eat very little. The larvae live under loose bark, where they prey on other invertebrates.

Dictyopterus simplicipes

Dictyopterus simplicipes

ADULT Wingcovers and legs red; pronotum red with black spot; thorax and abdomen black. **BODY LENGTH** 11 mm. **FOOD** Adult: Unknown. Larva: Invertebrates. **FOUND** From British Columbia to California. Look for adults in early spring.

BLISTER BEETLES Family Meloidae

The beetles in this family are usually narrow and elongate with soft wingcovers, a pronotum that is narrower than the head or wingcovers, and a narrow neck. This family gets its common name from the fact that some species, when threatened, secrete a chemical, called *cantharadin*, that causes human skin to blister. This chemical is used as a medicinal drug. The adults are commonly seen on plants, feeding on flowers and leaves. The life cycle of blister beetles is complex: the larval stages appear in several forms. The first larval instar is slender, long-legged, and active. Once the larva locates a host, it goes through more instars which progressively appear more grublike until it pupates. The larvae of most species are considered to be beneficial because they feed on grasshopper eggs. Other species attack bee larvae. The adults of some species are important crop pests.

Epicauta puncticollis

PUNCTATE BLISTER BEETLE

ADULT Shiny black and covered with short hairs. **BODY LENGTH** 15 mm. **FOOD** Adult: Plants (flowers and foliage). Larva: Grasshopper eggs. **FOUND** Throughout the region. Adult occurs in such diverse habitats as mountain meadows and weed-infested city lots.

Large numbers of adults may completely strip the flowers off plants such as the chicory *Cichorium intybus*, a common roadside weed. This species is so common that it is a wonder that any grasshopper egg can survive.

Epicauta puncticollis

Nemognatha lurida

Nemognatha lurida

ADULT Wingcovers orangish brown, often with black margins; pronotum and head orangish brown; legs, eyes, and antennae black. **BODY LENGTH** 10 mm. **FOOD** Adult: Plants (flowers). Larva: Native bees (parasitic). **FOUND** Throughout the region. Adult commonly found, as are native bees, on the pussypaws *Calyptridium umbellatum* and wild buckwheat (*Eriogonum* spp.).

■ SOFT-WINGED FLOWER BEETLES Family Melyridae

Although the members of this family are quite common, little is known about them. The adults are small, elongate-oval, and soft-bodied; many are brightly colored. Although the adults are predaceous, many are commonly found on flowers, feeding on pollen. Some are known to feed on the eggs of other insects. The larvae are predaceous.

Collops bipunctatus

TWO-SPOTTED SOFT-WINGED FLOWER BEETLE

ADULT Wingcovers iridescent dark blue; pronotum orange with two black spots. **BODY LENGTH** 5 mm. **FOOD** Adult and larva: Invertebrates. **FOUND** Throughout the region.

Collops bipunctatus

Collops versatilis

ADULT Wingcovers red, each with small, black spot at base and large, black spot near tip; male antennae with enlarged basal segments. **BODY LENGTH** 5 mm. **FOOD** Adult and larva: Invertebrates. **FOUND** Throughout the region. Adult found on flowers.

The male may possibly use its enlarged antennae to hold the female during mating.

Collops versatilis, male

POLLEN-FEEDING BEETLES Family Oedemeridae

The beetles in this small family are often mistaken for long-horned beetles (Cerambycidae). They are medium to large in size, have slender, soft bodies, and are usually pale tan to light red, but also black. The pronotum is somewhat narrowed posteriorly and narrower than the wingcovers. The antennae are long. The adults are usually found on flowers or foliage or under bark and are attracted to lights at night. They are pollen feeders. The larvae live in soil, usually where it is moist, or in decaying wood.

Ditylus quadricollis

Ditylus quadricollis

ADULT Black; pronotum narrowed posteriorly and narrower than wingcovers; wingcovers leathery; antennae long. **BODY LENGTH** 24 mm. **FOOD** Adult: Flowering plants (pollen). Larva: Rotting wood. **FOUND** Throughout the region. Adults common in spring around downed, rotting trees.

BARK-GNAWING BEETLES Family Ostomidae (Trogossitidae)

The beetles in this family are separated into two very different-looking groups. In one group, the body is elongate, slender, cylindrical, slightly flattened, and either green, blue, or black; the head is almost as wide as the pronotum; and the pronotum is widely separated from the base of the wingcovers. In the other group, the body is oval, flattened, and usually colored in shades of brown; the head is only about half as wide as the pronotum; and the pronotum is more closely joined to the base of the wingcovers. The elongated forms are predaceous; it is not known what the oval forms eat, but some species may feed on fungus. The adults and larvae are usually found under bark, in woody fungi, and in dry plant material. The larvae are predaceous, hunting their prey under bark, in the tunnels of wood-boring insects, or in fungi. A few species are considered to be pests because they also feed on stored grains.

Ostoma pippingskoeldi

Ostoma pippingskoeldi

ADULT Wingcovers and pronotum brown with black markings and flat margins. **BODY LENGTH** 9 mm. **FOOD** Adult: Unknown, but probably fungus. Larva: Invertebrates. **FOUND** Throughout the region, under bark.

Temnoscheila chlorodia

GREEN OSTOMID

ADULT Iridescent green and dark blue; head large with large jaws. **BODY LENGTH** 12 mm. **FOOD** Adult and larva: Insects. **FOUND** Throughout the region, under bark.

Temnoscheila chlorodia

■ GLOWWORM BEETLES Family Phengodidae

This small family of relatively uncommon beetles is closely related to firefly beetles (Lampyridae). The adult males are medium in size and have wingcovers that are shorter than the abdomen; most species are broad and flat, and some have antennae that are more feathery than those of the firefly beetles. All known females are luminescent and larviform, like those in some species of Lampyridae. The adults are found on foliage or on the ground. They often do not feed. The larvae live in leaf litter or under objects on the ground, preying on other invertebrates.

Zarhipis integripennis

WESTERN BANDED GLOWWORM

ADULT Male, wingcovers dark brown; pronotum, abdomen, and legs brownish orange; eyes large, globular, and dark brown; antennae large and finely branched. Female (and larva), creamy white (these areas glow) with dark brown dorsal bands. Male very similar to *Zarhipis tiemanni* except antennae much longer; female very similar to *Z. tiemanni.* **BODY LENGTH** Male 15 mm, female 65 mm. **FOOD** Adult: Probably none. Larva: Millipedes. **FOUND** Throughout the region.

Zarhipis integripennis, female

A resting male during the spring season can often be found slowly fanning the air with its antennae, trying to detect pheromones released by the females. The process by which a larva eats millipedes, which have heavily armored bodies, is as follows: the larva first paralyzes the millipede, using its jaws to sever the main nerve at the base of the millipede's head. It then injects a chemical to liquefy or soften the millipede's flesh so that it can ingest the millipede's body contents through its hollow jaws. The larva removes the millipede's head and proceeds to feed on the contents of the first segment of the millipede's body. It removes the "ring" that surrounded the segment and proceeds to the next segment. It continues in this manner until the millipede is consumed, and what remains of the millipede is a pile of disconnected rings.

Zarhipis tiemanni, male and female

Zarhipis tiemanni

TIEMANN'S GLOWWORM

ADULT Male, wingcovers dark brown; pronotum, abdomen, and legs brownish orange; eyes large, globular, and dark brown; antennae large and finely branched. Female (and larva), creamy white (these areas glow) with dark brown dorsal bands. Male very similar to *Zarhipis integripennis* except antennae much shorter; female very similar to *Z. integripennis*. **BODY LENGTH** Male 14 mm, female 65 mm. **FOOD** Adult: None. Larva: Invertebrates. **FOUND** There is very little information on the range of this species (Pete has collected it in Oregon and California).

A resting male during the spring season can often be found slowly fanning the air with its antennae, trying to detect pheromones released by the females.

■ LAMELLICORN BEETLES Family Scarabaeidae

This very large family is easily recognized by the adults' antennae, the last three segments of which are expanded into platelike structures, called *lamellae*. These beetles are generally robust and convex in shape but vary greatly in size. They vary greatly in color, from drab brown or black to iridescent blue or green. The adults of many species are important scavengers that recycle dung or decaying organic matter; others feed on foliage, fruits, and flowers; some live in the nests or burrows of vertebrates or in the nests of termites or ants; a few feed on fungi. The larvae are white and C-shaped and have a brown head with prominent jaws. Many live in the soil and feed on roots. Others live in dung or carrion, under bark, in fungi, or in the nests of burrowing vertebrates and social insects.

Cremastocheilus armatus

Cremastocheilus armatus

ANT SCARAB

ADULT Flattened, hard-bodied, and black; wingcovers, pronotum, and head covered with punctures. **BODY LENGTH** 15 mm. **FOOD** Adult: Ant larvae. Larva: Material in ant nests. **FOUND** Throughout the region, in the nests of various species of formicid ants.

Ants have been observed on this species, gnawing at the pubescent glandular areas located on the sides of the beetle's thorax. These glandular areas apparently provide the ants with food. This beetle is quick to play dead if threatened.

Dichelonyx spp.

ADULT Wingcovers iridescent green; pronotum black. Can be identified to species only by internal structures. **BODY LENGTH** 13 mm. **FOOD** Adult: Conifers (needles). Larva: Plants (roots). **FOUND** Throughout the region. Look for adults in early spring.

Dichelonyx sp.

Euphoria indus

BUMBLE FLOWER BEETLE

ADULT Bumble bee mimic—wingcovers brown with light brown longitudinal streaks and black speckling; pronotum and abdomen covered with dense, light brown hairs. Compare to *Lichnanthe rathvoni*. **BODY LENGTH** 16 mm. **FOOD** Adult: Rotting fruit and nectar (Pete has also collected them feeding on thistle pollen). Larva: Rotting wood, dung, and humus. **FOUND** Throughout the region, more commonly inland than on the coast.

Euphoria indus

Hoplia dispar

HOPLIA

ADULT Brown to black; pronotum and wingcovers covered with orangish brown to light brownish green scales. **BODY LENGTH** 8 mm. **LARVA** Maggotlike. **FOOD** Adult: Plants (often pollen, but also leaves and flowers). Larva: Plants (roots). **FOUND** Oregon and California. Adults common from spring to early summer on flowers in the mountains.

Hoplia dispar

Lichnanthe rathvoni

RATHVON'S SCARAB

ADULT Bumble bee mimic—wingcovers small, dark brown with small patches of light yellow hairs; abdomen with wide bands of long, light yellow to pale orange hairs and narrow bands of short, black hairs. Compare to *Euphoria indus*. **BODY LENGTH** 15 mm. **FOOD** Adult: Plants (pollen on flowers). Larva: Probably plants (underground roots) and de-

Lichnanthe rathvoni

Onthophagus taurus

Polyphylla decemlineata, male

Polyphylla decemlineata, larva

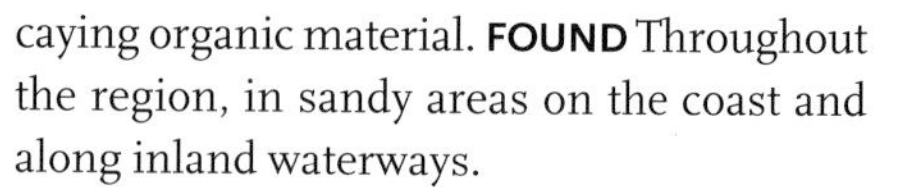

caying organic material. **FOUND** Throughout the region, in sandy areas on the coast and along inland waterways.

Some entomologists place this species in the family Glaphyridae. The adult is a slow, clumsy flyer (with a flight pattern very similar to a bumble bee's) and can often be caught by hand.

Onthophagus taurus

ADULT Black; head with two long, swept-back horns. **BODY LENGTH** 9 mm. **FOOD** Adult and larva: Dung. **FOUND** Throughout the U.S. Pete has found adults in fresh bear dung.

Polyphylla decemlineata

TEN-LINED JUNE BEETLE

ADULT Wingcovers and pronotum brown with white longitudinal lines; male antennae distinctly large and flattened. **BODY LENGTH** 35 mm. **LARVA LENGTH** Up to 50 mm. **FOOD** Adult: Conifers. Larva: Plants (roots). **FOUND** Very common throughout the region.

The life cycle of this species is three to four years. When the adult is handled, it makes a sound like a child's cry.

BARK BEETLES Family Scolytidae

The members of this family are small, cylindrical beetles that carve elaborate tunnels in the inner bark or wood of trees. They are usually black or brown in color and have elbowed, distinctly clubbed antennae. The family comprises two groups: the bark, or engraver, beetles, and the ambrosia, or timber, beetles. The bark beetle adults and larvae live beneath the bark of trees, where they excavate branching tunnels and eat wood. These tunnel patterns, or "engravings," are distinct for each species. When a bark beetle female bores into a live tree to lay her eggs, the tree may respond by producing copious amounts of pitch at the site of her entry hole. If a beetle bores into a healthy tree, the tree can often force the beetle out of its hole with pitch, thus protecting itself from infestation. The ambrosia beetles tunnel into the heartwood and feed on the fungi that they cultivate in their tunnels. Each species usually feeds on one particular type of fungus. When the females emerge from their natal tree and fly to another tree, they carry the fungus on their bodies and consequently inoculate the new host tree with the fungus. Most scolytids infest conifers and are usually host specific. They may cause considerable damage but usually infest weakened, dying, or dead trees or recently cut timber. The damage is mainly caused by either the beetles' feeding on the wood (bark beetles) or spreading disease from one tree to another (ambrosia beetles).

Dendroctonus valens

RED TURPENTINE BEETLE

ADULT Reddish brown. **BODY LENGTH** 9 mm. **FOOD** Adult and larva: Pine (*Pinus* spp.) (wood borer). **FOUND** Throughout the region. Look for presence of large, reddish brown *pitch tube* (a glob of pitch and sawdust), which indicates a beetle's entrance hole, usually near base of host tree.

This is the largest bark beetle in the region. One of the common names for this species is *barber beetle* because of its supposed ability to clip hair with its powerful jaws.

Dendroctonus valens

CARRION BEETLES Family Silphidae

The beetles in this family have large, flattened bodies, clubbed antennae, and wingcovers that often leave part of the abdomen exposed. Most are black or dark brown, some with bright red or orange markings. These common beetles are often associated with carrion (hence the common name), and to a lesser extent decaying vegetation, which the adults locate with the help of receptors on their antennae. Most adults and larvae feed on the carcasses, which some species bury. The females of some species also feed their larvae regurgitated food. A few species prey on maggots and other animals that occur in decaying organic matter. Some species occur in fungi, and a few occur in ant nests. Some larvae prey on snails; a few are plant feeders. The adults are protected from vertebrate predators by producing a foul-smelling liquid from their anus. It is unknown if this substance is toxic or simply has an unpleasant taste.

Heterosilpha ramosa

Heterosilpha ramosa, larva

Nicrophorus defodiens

Heterosilpha ramosa

GARDEN SILPHID

ADULT Black; wingcovers, each with three irregular, longitudinal ridges. **BODY LENGTH** 16 mm. **LARVA** Black and covered with overlapping dorsal plates. **FOOD** Adult and larva: Dead or decaying animal or plant material. **FOUND** Throughout the region.

Also known as *Silpha ramosa*.

Nicrophorus defodiens

ADULT Black; wingcovers with yellowish orange spots or bars. **BODY LENGTH** 17 mm. **FOOD** Adult and larva: Dead animals. **FOUND** Throughout the region.

This species is probably the most common

carrion beetle in the region. Although considered to be "burying beetles," the adults do not bury dead animals, but instead cover them with leaf litter.

Nicrophorus nigritus

BLACK BURYING BEETLE

ADULT Black; antennae with orange, sometimes light orange, tips. **BODY LENGTH** 18 mm. **FOOD** Adult and larva: Dead animals. **FOUND** Throughout the region.

In order to produce another generation, the adults pair up and search for recently deceased animals. When they find a body, they bury it and strip off the animal's fur, if it has any. In the process, they cover the body with their saliva, which chemically protects the body from fly eggs, bacteria, and fungi and may kill any fly larvae that might be present on the body. The female then lays her eggs on the body, and from these emerge the larvae, which feed on the body. The adult female shown in the photograph is guarding the entrance to a hole that houses a dead mouse that she and her mate had buried and are raising a brood on. Also, the adults usually have mites on their bodies, with whom they share a symbiotic relationship: the beetles carry the mites to the carcasses, where the mites dine on fly maggots and any other potential competitors of the beetle larvae.

Nicrophorus nigritus

■ ROVE BEETLES Family Staphylinidae

This is one of the largest families of beetles, with over 3000 species in North America. The adults can usually be recognized by their very short wingcovers. Most adults have slender, elongate bodies that are black or brown. Body size varies considerably, but most are small. The jaws are very long, slender, and sharp; larger beetles can inflict a painful bite when handled. The adults are very active and run or fly rapidly. When running, they frequently raise the tip of the abdomen, much like a scorpion. The adults and larvae occur in a variety of habitats, but they are probably most often found on decaying material, especially dung or carrion. It is thought that the adults and larvae of most species prey on invertebrates.

Ocypus olens

DEVIL'S COACH HORSE

ADULT Black. **BODY LENGTH** 30–33 mm. **FOOD** Adult and larva: Slugs and snails. **FOUND** Throughout most of the U.S., commonly in gardens.

Also known as *Staphylinus olens*. This spe-

Ocypus olens

cies was introduced from Europe. It seems to be more common, or at least more active, in the fall after the rains start. When threatened, the adult raises its rear end, and two creamy white glands emerge from the tip of the abdomen and release a smelly substance.

Thinopinus pictus

Thinopinus pictus

PICTURED ROVE BEETLE

ADULT Wingcovers grayish white, each with curved black band; hindwings absent; pronotum light brown with two black rings; head light brown with two black bands; abdomen grayish white with black bands. **BODY LENGTH** 20 mm. **FOOD** Adult and larva: Invertebrates. **FOUND** Throughout the region, along ocean beaches. Look for holes in the sand above the waterline; beetle will be 3–9 inches below the surface.

■ DARKLING BEETLES Family Tenebrionidae

This is the fifth largest family of beetles in North America. Although many species are common, the family is morphologically very diverse, and, since the features used to separate genera and species are hard to distinguish, it is not well represented in the literature. Most of these medium- to large-sized beetles are hard-bodied and black or dull brown and may be confused with the ground beetles (Carabidae). The antennae are often clubbed. The adults and larvae are found in almost every type of habitat except water; many species inhabit very dry environments. The adults and larvae feed on plant material. A few are common pests of stored grain and flour.

Eleodes scabrosa

Eleodes scabrosa

ADULT Black; head and pronotum covered with tiny pits; wingcovers fused, covered with pits larger than those on head and pronotum; hindwings absent. **BODY LENGTH** 10 mm. **FOOD** Adult and larva: Living and dead plants. **FOUND** Throughout the region, along ocean beaches. Look for adults in spring and early summer on sand or under debris on beach.

The sexual morphology of the males is the criterion used in identification keys to identify to species.

■ IRONCLAD BEETLES Family Zopheridae

The species in this family are sometimes placed in the darkling beetle family (Tenebrionidae). The adults are medium-sized, very hard-bodied, brown beetles that resemble a piece of bark. They are found under bark or on the fruiting bodies (conks and mushrooms) of fungi. Little is known about their biology.

Phellopsis obcordatus

ADULT Flattened; wingcovers pitted and grooved. **BODY LENGTH** 15 mm. **FOOD** Adult: Fungus (fruiting bodies). **FOUND** Throughout the region. Look for adults on trees with conks or mushrooms growing on them.

Phellopsis obcordatus

FLIES Order Diptera

Diptera (Greek, *di*, two; *ptera*, wings) is the fourth largest of the insect orders, with approximately 98,500 species described and over 16,900 of those species occurring in the United States and Canada. This order is found worldwide and is abundant in individuals as well as species almost everywhere.

As the scientific name of the order implies, a major feature of this order that makes flies easily recognizable is the possession of only one pair of wings (forewings). The second pair (hindwings) has either been reduced to small, knobbed balancing organs, called *halteres*, or is completely absent. Most flies are also distinguishable from other insects by the fact that the membranous forewings are usually translucent with relatively few veins.

Flies are minute to medium in size and have a body that is usually soft or leathery, very uniform in shape, and usually dull in color. In most species, the mouthparts have been modified for feeding on liquids by piercing and sucking, lapping, or sponging. The similarity of body shapes and drab colors makes it difficult to identify flies to species.

Flies undergo complete metamorphosis. The larvae, often called *maggots*, are usually legless and wormlike.

Flies are prevalent in most types of habitats, and the adults and larvae are either plant feeders, predators, scavengers, or parasites. The adults are often the first flying insects to emerge in late winter and early spring. They are usually active during the day and often found around flowers and other vegetation, refuse, feces, or animals (dead or alive). They feed on a variety of plant and animal fluids; most species feed on nectar and some on sap, but many others suck blood, with many predaceous on other insects. The larvae of many species live in water, but others live in soil, in decaying plant or animal matter, or as parasites of both vertebrates and invertebrates. Plant-feeding larvae commonly live within the plant tissue, either as leaf miners, stem or root borers, or gall insects.

Flies are generally regarded as having a negative effect on animal populations, especially those of humans, because they are responsible for much loss of life as well as economic loss. One reason for this reputation is that flies are the most important group of insects in the transmission of human and agricultural animal diseases. Approximately 50 percent of the Earth's human population has diseases caused by pathogens or parasites spread by flies. They are the vectors of such diseases as malaria, typhoid, sleeping sickness, yellow fever, and dysentery. Another reason for the flies' infamous reputation is that some species are important agricultural pests, causing significant crop losses.

Unfortunately, flies are much less appreciated for their other roles in natural processes. They are important in the pollination of plants and, as scavengers, in the decomposition of dead plant and animal matter. Their role in the recycling of organic material is exceeded only by that of bacteria and fungi. They are also a major source of food for wildlife. Some species are extensively used for biological control on insect pests of crops and others for the control of disease vectors.

Key to Diptera Families

ROBBER FLIES
Family Asilidae, page 88

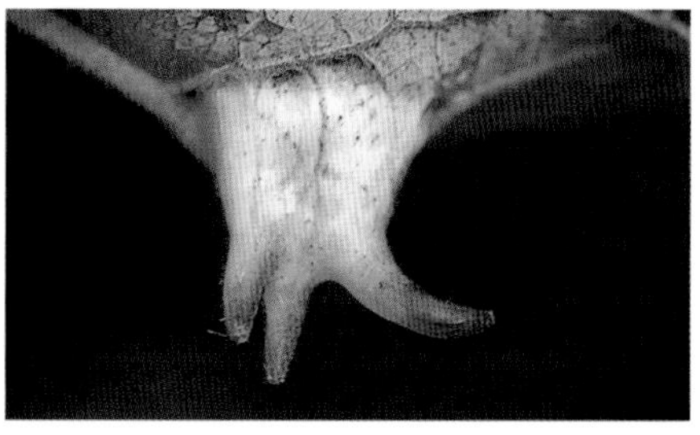

GALL MIDGES
Family Cecidomyiidae, page 88

HOUSE FLIES
Family Muscidae, page 90

PHANTOM CRANE FLIES
Family Ptychopteridae, page 90

FLOWER FLIES
Family Syrphidae, page 91

PARASITIC FLIES
Family Tachinidae, page 91

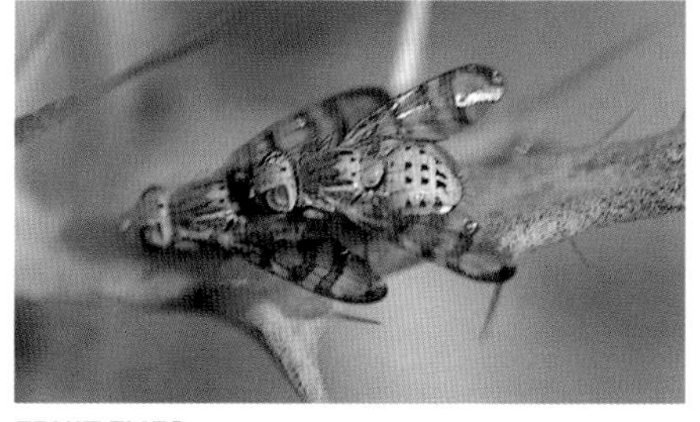

FRUIT FLIES
Family Tephritidae, page 92

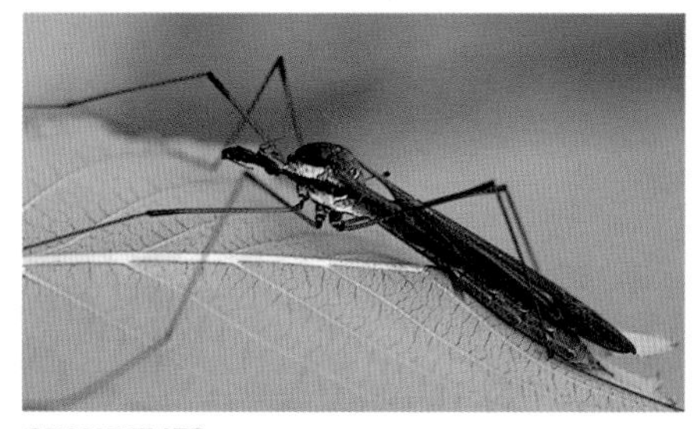

CRANE FLIES
Family Tipulidae, page 92

ROBBER FLIES Family Asilidae

This is a large family, with some species quite common. These flies are medium to large in size and have bulging eyes. Many species have a long, slender body with few hairs; others are short, stout, and very hairy and resemble bees. The adults prey on a variety of insects. They can inflict a painful bite when handled. The larvae feed on other insect larvae in decaying wood or loose soil or under bark or fallen leaves.

Laphria sackeni

Laphria sackeni

SACKEN'S ROBBER FLY

ADULT Bumble bee mimic—black with yellow hairs on face, thorax, tibia, and tip of abdomen. **BODY LENGTH** 22 mm. **FOOD** Adult: Insects. Larva: Insect larvae. **FOUND** Oregon and California.

GALL MIDGES Family Cecidomyiidae

The adult members of this very large family are tiny, delicate flies that have long legs and usually long antennae. The larvae are tiny maggots and in many species are brightly colored. More than half of the species cause galls on plants; the form of the gall is usually characteristic of the species. Most of the other species feed on plants, although a few common ones prey on aphids. Gall-forming species have a complex relationship with the plants they invade. Other insect species may invade the gall and take over as a parasite or predator on the larva(e) inside. In some genera, the larvae undergo a process called *paedogenesis* in which the larvae asexually reproduce themselves. In this process, a mature (mother) larva produces daughter larvae within itself. The daughter larvae eventually eat and emerge from the mother larva. They in turn may produce more larvae in a similar fashion through several generations. The last generation of larvae pupate.

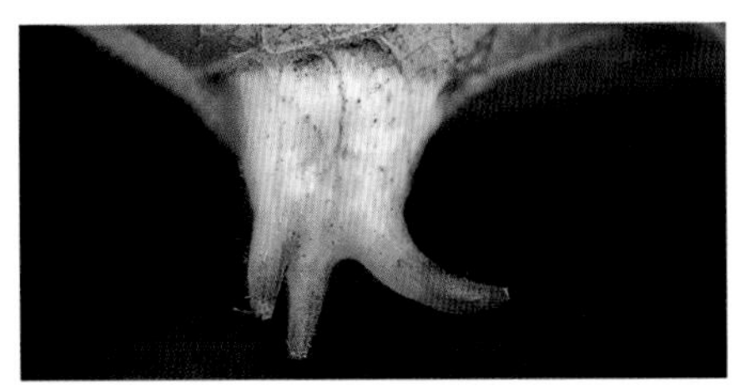

Cecidomyia salicisverruca, galls

Cecidomyia salicisverruca

WILLOW LEAF GALL MIDGE

GALL Cone-shaped; yellowish green, red when exposed to sun; protrudes from bottom of leaf; when in groups of two or more, often coalesce (three galls shown). **GALL HEIGHT** 5 mm. **FOOD** Larva: Coastal willow (*Salix hookeriana*), shining willow (*S. lucida*), and other willows. **FOUND** California, but Pete suspects that it occurs throughout the region (he has collected galls in Oregon and California).

Also known as *Iteomyia salicisverruca*.

Paradiplosis tumifex

BALSAM FIR GALL MIDGE

GALL Enlarged area on fir needle, which is often distorted as a result. **GALL LENGTH** 2 mm. **FOOD** Larva: Grand fir (*Abies grandis*), white fir (*A. concolor*). **FOUND** Throughout much of the U.S.

Look for the gall in the fall when it is at its largest. This species has an interesting relationship with the midge *Dasineura balsamicola*, which, since it cannot form its own gall, is completely dependent upon *Paradiplosis tumifex*. The *P. tumifex* female lays one egg on a fir needle. After emerging from the egg, the larva feeds on the needle, which starts to enclose the larva in a gall. Before the gall completely forms, a *D. balsamicola* female lays her egg near the *P. tumifex* larva. When the *D. balsamicola* larva emerges, it immediately searches for the *P. tumifex* larva and lies next to it. The gall then encloses both larvae. The only larva to emerge from the gall is the *D. balsamicola*, since it grows faster than the *P. tumifex* larva. An animal, such as *D. balsamicola*, that lives in the home of another is called an *inquiline*.

Paradiplosis tumifex, gall

Rabdophaga salicisbatatus, gall

Rhopalomyia californica, fresh (green) gall and previous year's gall

Rabdophaga salicisbatatus

WILLOW POTATO GALL

GALL Long, irregular, potatolike shape; on willow stems. **GALL LENGTH** 40 mm. **FOOD** Larva: Coastal willow (*Salix hookeriana*), shining willow (*S. lucida*), and other willows (*Salix* spp.). **FOUND** Throughout the region.

Other species of this genus that occur in the region have willow galls that are shaped differently from that of *Rabdophaga salicisbatatus*.

Rhopalomyia californica

COYOTE BRUSH GALL MIDGE

GALL Globular; yellowish green, red when exposed to sun; on tips of branches in the spring. **GALL WIDTH** 20 mm. **FOOD** Larva: Coyote brush (*Baccharis pilularis*). **FOUND** Oregon and California.

HOUSE FLIES Family Muscidae (Scathophagidae)

The members of this large family closely resemble its most familiar member, the house fly. The adults of some species prey on other insects, but most eat and breed in feces and decaying matter. Consequently, muscid flies are important vectors of bacteria-caused diseases, although only a few species, such as the tse tse fly, spread disease directly by biting humans and other animals. The larvae feed on living plants as well as decaying matter and the inhabitants thereof.

Scathophaga stercoraria, adult feeding on leafhopper

Scathophaga stercoraria

GOLDEN-HAIRED DUNG FLY

ADULT Thorax and abdomen yellowish brown to shiny golden brown and densely covered with short, golden yellow hairs. **BODY LENGTH** 8 mm. **FOOD** Adult: Invertebrates. Larva: Probably other insects in dung. **FOUND** Throughout the U.S., commonly in cow pastures and in decaying vegetation in wet areas.

The adults are one of the first insects to emerge in early spring.

PHANTOM CRANE FLIES Family Ptychopteridae

The adult members of this family resemble crane flies (Tipulidae). They have a long, slender, delicate body and long, slender legs, which in some species, are conspicuously banded with white. The adults are found in shady woods or other damp places and probably do not feed. The larvae feed on decaying vegetation in stagnant or nearly stagnant water.

Bittacomorpha occidentalis

Bittacomorpha occidentalis

ADULT Black; legs very long, delicate, and with white bands. **BODY LENGTH** 13 mm. **FOOD** Larva: Decaying plant matter. **FOUND** Throughout the region, along streams in shade.

The adult seems to float rather than fly, like a large tuft of dispersing willow seed in the springtime, and, when it flies in the shade, only the white leg bands are visible, making the insect seem to disappear.

FLOWER FLIES Family Syrphidae

This is a large family, and many of its species are conspicuously abundant. The adults of most species are brightly colored with bands of black and yellow and resemble various bees or wasps. They are harmless to humans and do not bite or sting. Many commonly hover at flowers. They feed on nectar and pollen, and many are pollinators. The larvae vary considerably in appearance and habits. Many are predaceous, especially on aphids, and some feed on decaying organic matter.

Eupeodes lapponica

ADULT Thorax bright golden brown; abdomen with golden yellow and black bands. **BODY LENGTH** 10 mm. **LARVA** Mottled with gray, black, and brown and tapers anteriorly. **FOOD** Larva: Invertebrates (Pete has found larvae feeding on European spruce aphids [*Elattobium abietinum*]). **FOUND** Throughout the region.

Eupeodes lapponica

PARASITIC FLIES Family Tachinidae

This is the second-largest family of flies in North America. Its members are found almost everywhere. The adults have a medium-sized, stout, bristly body with a prominent thorax. The females of most species lay their eggs on the bodies of other invertebrates. It is not uncommon to see a caterpillar with tachinid eggs cemented to its skin. In other species, the females lay their eggs on plants that are food to the host invertebrates or inject the larvae into the host species. The larvae then feed within the host, eventually killing it. When fully developed, the larvae of most species leave the host and pupate nearby in the ground (some pupate within the host). These flies are a major factor in the regulation of populations of many invertebrate species.

Epalpus signifer

ADULT Abdomen shiny black, covered with stiff, black hairs, and with large, golden yellow spot on posterior. **BODY LENGTH** 9 mm. **FOOD** Larva: Lepidoptera larvae (parasitic). **FOUND** Throughout the region. Look for adults on flowers.

Epalpus signifer

FRUIT FLIES Family Tephritidae

This large family consists of small- to medium-sized flies, most of which have complex patterns on their wings (these species are commonly called *picture-winged flies*). The members of this family are sometimes also called *peacock flies* because of their habit of rhythmically waving their wings up and down, like a peacock moving its tail feathers. The adults are found on flowers or other vegetation. The larvae of many species feed on the fruits or seeds of plants. Some are serious pests in orchards, but others are used to control weeds such as yellow star-thistle.

Orellia occidentalis, mating pair

Orellia occidentalis

ADULT Creamy white; thorax with large, gray patch; abdomen with small, dark brown spots; wings patterned; eyes green. **BODY LENGTH** 4 mm. **FOOD** Larva: Swamp thistle (*Cirsium douglasii*) and other thistles (*Cirsium* spp.) (flower buds). **FOUND** Throughout the region.

The female lays her eggs in flower buds. The larvae that emerge from these eggs feed inside the buds and open flowers, often destroying all the seeds within.

CRANE FLIES Family Tipulidae

This family is the largest in the order in the United States and Canada. Many members are common and abundant. Most of these gray or brown flies resemble overgrown mosquitoes with very long, delicate legs, which break off easily. The adults are common in a great variety of habitats but occur mainly in moist situations where there is abundant vegetation. Most do not feed; some may feed on nectar. The larvae of most species are aquatic or semi-aquatic and feed on decaying organic matter; some feed on living plants.

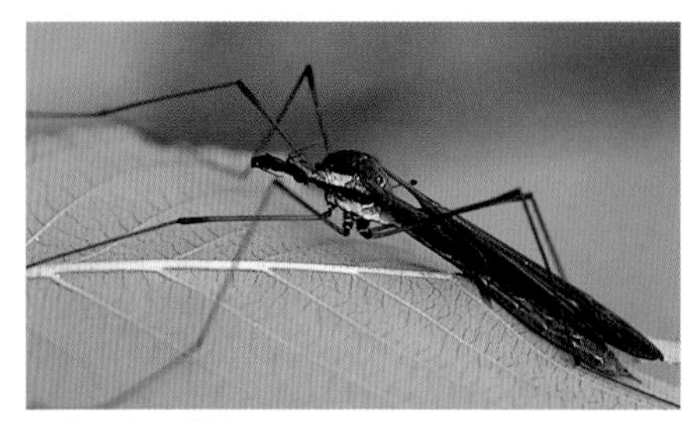

Holorusia rubiginosa

Holorusia rubiginosa

GIANT CRANE FLY

ADULT Reddish brown; thorax with broad, white bands on sides. **BODY LENGTH** 34 mm. **FOOD** Larva: Decaying plant material. **FOUND** Throughout the region, around streams or other wet areas.

Also known as *Holorusia rubinosa* and *H. hespera*. This species is one of the largest flies in the world. Another common name for this fly is *mosquito hawk*, even though the adult is not predaceous.

Tipula paludosa

Tipula paludosa

EUROPEAN CRANE FLY

ADULT Thorax brown; abdomen light brown. **BODY LENGTH** 30 mm. **LARVA** Large; leathery; light brown. **FOOD** Larva: Grasses. **FOUND** Throughout the region, very commonly in pastures and lawns.

This European species was accidentally introduced into British Columbia in 1965 and has since spread south to California, where it was first collected in 1999. The mature larvae emerge from the ground and wiggle their way around looking for a place to pupate, often ending up on sidewalks in large numbers.

Tipula paludosa, bucket of larvae

Hemiptera (Greek, *hemi*, half; *ptera*, wings) is a large and widely distributed order that comprises approximately 50,000 species worldwide, with over 3500 recorded in the United States and Canada. It is abundant in numbers of individuals as well as species in both temperate and tropical regions. Hemipterans are also commonly referred to as *true bugs* or *plant bugs*. Some entomologists combine Hemiptera and Homoptera into one order, Heteroptera.

The adults range in size from tiny to large, but most in the Pacific Northwest region are small to medium. They occur in a wide variety of colors, but most are dark. The majority of species have two pairs of well-developed wings, although a few species are short-winged or wingless. One of the most distinctive characteristics of this order, and from which the order derives its name, is the structure of the forewings. In most of the Hemiptera, the basal part of the forewing is thickened and leathery, and the apical part is membranous and usually translucent. The hindwings, the flying wings, are completely membranous and slightly shorter than the forewings. When a bug is at rest, its wings are held flat over its abdomen, with the membranous portions of the forewings overlapping and the hindwings folded under the forewings. Another distinctive feature of this order is the scutellum, a triangular portion of the thorax that is exposed between the bases of the wings. The mouthparts are elongated into a beak that is used for piercing and sucking and usually extends backward between the legs when not in use. Many terrestrial species emit a distinctive odor, particularly when they are disturbed. This odor is thought to repel potential predators.

Bugs undergo incomplete metamorphosis. The immature bugs, called *nymphs*, closely resemble the adults except that they are wingless.

The majority of bug species are terrestrial, but many others are aquatic or semi-aquatic. Most terrestrial species feed on plant fluids and can be commonly found in open fields, particularly on grasses. A few terrestrial species are predaceous, with some feeding on the blood or other body fluids of mammals and birds. Almost all the aquatic species prey on other invertebrates and small vertebrates.

Although some bug species are major crop pests, others that prey on insects help keep down populations of insect species that are harmful to crops. The bugs that are bloodsucking parasites are capable of transmitting diseases of animals, including humans.

Key to Hemiptera Families

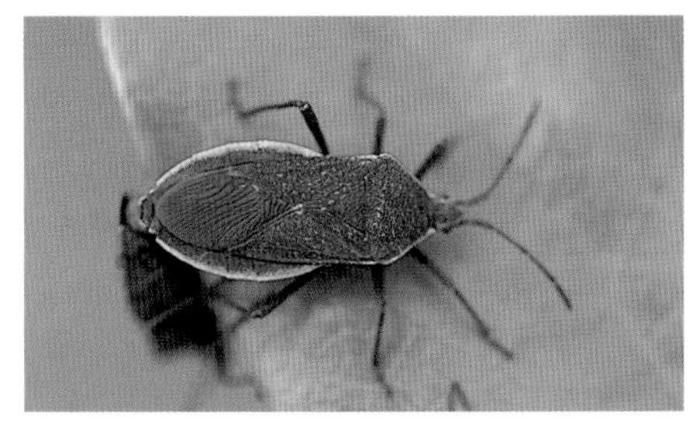

SQUASH BUGS
Family Coreidae, page 96

TOAD BUGS
Family Gelastocoridae, page 97

RED BUGS
Family Largidae, page 97

SEED BUGS
Family Lygaeidae, page 98

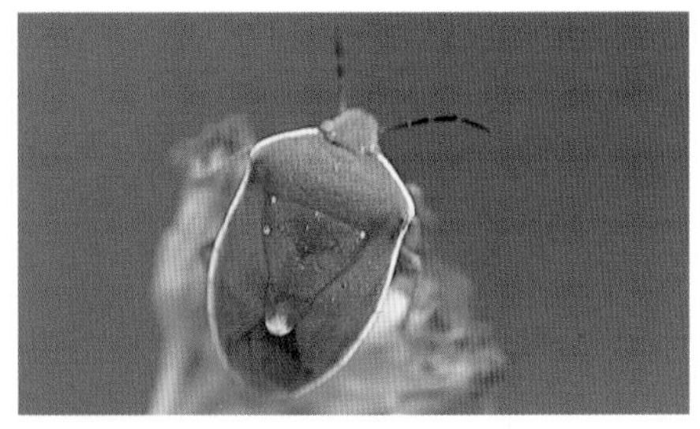

STINK BUGS
Family Pentatomidae, page 100

AMBUSH BUGS
Family Phymatidae, page 105

ASSASSIN BUGS
Family Reduviidae, page 106

SCENTLESS PLANT BUGS
Family Rhopalidae, page 107

SQUASH BUGS Family Coreidae

The members of this large family are medium- to large-sized with the membrane of their front wings having numerous veins. The body is usually brown or gray. Some species have hind legs that are enlarged and flattened and appear leaflike. All species feed on plants and have scent glands that can secrete an unpleasant odor when disturbed. These bugs are usually inconspicuous unless one dislodges them from the vegetation with a sweep net or by beating the vegetation.

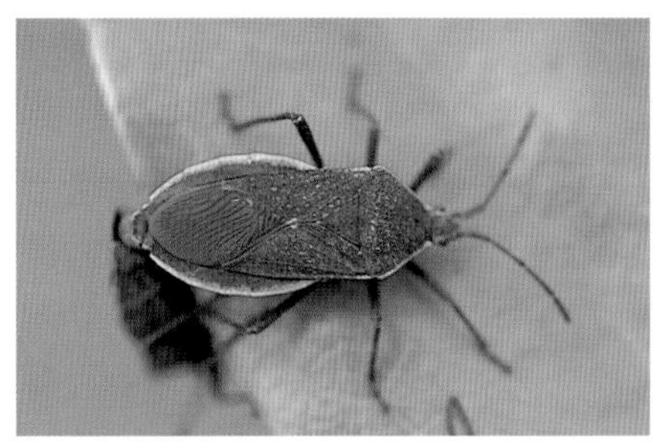

Anasa tristis

Anasa tristis, nymphs

Leptoglossus occidentalis

Anasa tristis

SQUASH BUG

ADULT Dark brown; pronotum with pale orange or pale brown lateral margins; abdomen with protruding margins edged with pale orange. **BODY LENGTH** 17 mm. **NYMPH** Brownish gray. **FOOD** Adult and nymph: Coast man-root (*Marah oreganus*) and other gourds (Cucurbitaceae). **FOUND** Throughout much of the U.S.

Leptoglossus occidentalis

WESTERN CONIFER SEED BUG

ADULT Brown; wings with yellow or white zigzag line through middle; hind tibia flattened like a leaf. **BODY LENGTH** 17 mm. **FOOD** Adult and nymph: Conifers (cones). **FOUND** Throughout the region. Common but rarely seen since it spends much of its life in the tree canopy.

TOAD BUGS Family Gelastocoridae

These bugs resemble small toads in appearance and locomotion. They have a short, broad body, large, bulging eyes, hind legs adapted for jumping, and short front legs that are modified to catch small insects. When disturbed, they hop away or crouch down. They usually occur along moist margins of streams and ponds. The eggs are laid in sand, and the adults often spend a part of their lives down in the sand.

Gelastocoris oculata

ADULT Mottled with gray, brown, and black. **BODY LENGTH** 9 mm. **FOOD** Adult and nymph: Invertebrates. **FOUND** Throughout most of the U.S., along banks of streams, rivers, and ponds.

When sitting on dry sand, this bug is almost impossible to see unless it moves.

Gelastocoris oculata

RED BUGS Family Largidae

These bugs, which usually occur in the southern states, are similar in appearance to seed bugs (Lygaeidae). They are medium- to large-sized, elongate-oval, and often brightly marked with black and red. They feed on plants, and some are crop and garden pests.

Largus cinctus

BORDERED PLANT BUG

ADULT Black; pronotum with orange posterior margin; wings with orange speckling and orange outer margin; abdomen bordered with orange. **BODY LENGTH** 13 mm. **NYMPH** Early stages, red; later stages, iridescent blue-black with red dorsal spot at base of abdomen. **FOOD** Adult and nymph: Plants. **FOUND** Oregon and California.

This species is one of the most conspicuous bugs in the region. Adults often occur in large numbers in the fall. On warm sunny days in winter, overwintering adults are often seen sunning themselves on a rock, tree, or the side of a building. In the spring it is quite common to see pairs of adults coupling.

Largus cinctus

SEED BUGS Family Lygaeidae

This is the second-largest family of bugs in North America, and many of its members are common. These small bugs have elongate or oval, hard bodies. Many species are conspicuously marked with bands or spots of red, black, or white. Most species, even those that have enlarged legs and appear raptorial, feed on seeds. A few feed on sap, and a few others are predaceous. The sweep of a net in tall grass will normally catch adults and nymphs in large numbers.

Lygaeus kalmii

Neacoryphus bicrucis

Lygaeus kalmii

SMALL MILKWEED BUG

ADULT Above, black with red X; wing membranes each with white spot. **BODY LENGTH** 10 mm. **NYMPH** Red; wing pads black. **FOOD** Adult and nymph: Showy milkweed (*Asclepias speciosa*) and other milkweeds (*Asclepias* spp.), as well as other plants (seeds and seed pods). **FOUND** Common throughout the region.

Pete has seen large populations of this bug on the sand verbena *Abronia latifolia* in the fall. This species and the large milkweed bug (*Oncopeltus fasciatus*) are often found feeding together on milkweed. *Lygaeus kalmii* is much more common than *O. fasciatus* and is usually the only milkweed bug found in coastal areas.

Neacoryphus bicrucis

ADULT Pronotum red with two black spots anteriorly; wings red and edged with white; scutellum and wing membrane black and edged with white. **BODY LENGTH** 9 mm. **FOOD** Adult and nymph: The groundsel *Senecio triangularis*. **FOUND** Range unknown but probably same as host.

Oncopeltus fasciatus

LARGE MILKWEED BUG

ADULT Dark brown to black; pronotum with red to orange lateral margins; wings each with two broad, red to orange bands; head with red to orange Y. **BODY LENGTH** 18 mm. **NYMPH** Red to orange; wing pads, legs, and antennae black. **FOOD** Adult and nymph: Milkweed (*Asclepias* spp.) (seeds and seed pods). **FOUND** Widespread throughout much of the U.S. Look for adults on seed pods of host plants.

The adults and nymphs are often found together in small to large groups on the host plant. This is the milkweed bug usually pictured in insect field guides and textbooks, although the small milkweed bug (*Lygaeus kalmii*) is much more common and is usually the only milkweed bug found in coastal areas. The two species are often found feeding on the same plant.

Oncopeltus fasciatus

Oncopeltus fasciatus, nymph

Xanthochilus saturnius

ADULT Above, pale brown with black diamond- and triangle-shaped areas. **BODY LENGTH** 7 mm. **FOOD** Adult and nymph: Plants. **FOUND** Throughout the region.

Xanthochilus saturnius

STINK BUGS Family Pentatomidae

The members of this large, well-known family get their common name from the odor they produce when threatened. This smell may not necessarily be as unpleasant to humans as the name implies. These medium- to large-sized bugs are easy to recognize by their shieldlike shape. They are brown or green in color, sometimes marked with bright red and other colors. They either feed on plants or prey on insects.

Acrosternum hilaris

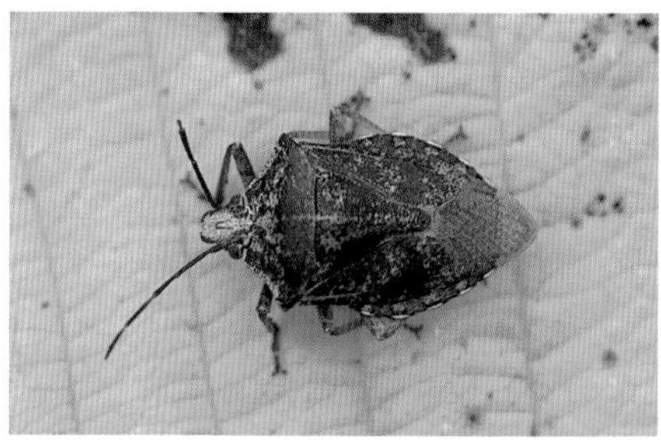
Apateticus crocatus

Apateticus crocatus, nymph feeding on *Nematus chalceus* larva

Acrosternum hilaris

GREEN STINK BUG

ADULT Bright green with yellow, orange, or red lateral margins; antennae with three black segments at tip. Broader and slightly longer than *Chlorochroa uhleri*. **BODY LENGTH** 16 mm. **FOOD** Adult and nymph: Plants. **FOUND** Common throughout the U.S.

Apateticus crocatus

YELLOW PREDACEOUS STINK BUG

ADULT Two common variations—dark brown mottled with pale brown and abdomen with mottled green protruding margins; or completely yellow. **BODY LENGTH** 13 mm. **NYMPH** Later instars black with red markings. **FOOD** Adult and nymph: Invertebrates. **FOUND** Throughout the region. Look for adults and nymphs on plants, such as willows, with heavy populations of leaf beetle larvae.

This species is not as common as the spined stink bug (*Podisus maculiventris*).

Banasa dimiata

ADULT Yellowish green; wings dark red; pronotum with dark red on posterior; scutellum with yellow tip. **BODY LENGTH** 10 mm. **FOOD** Adult and nymph: Plants. **FOUND** Throughout much of the U.S.

Pete has found adults and nymphs on the leaves and fruits of native gooseberry or currant (*Ribes* spp.) and on the fruits of wax myrtle (*Myrica californica*).

Banasa dimiata

Chlorochroa ligata

CONCHUELA BUG

ADULT Dark green; scutellum with red, orange, or white tip; pronotum and abdomen bordered with red, orange, or white. **BODY LENGTH** 16 mm. **FOOD** Adult and nymph: Plants. **FOUND** Throughout the region. Look for adults on flowers in late summer and fall.

This species and *Chlorochroa uhleri* are two of the most common large stink bugs in the region.

Chlorochroa ligata

Chlorochroa sayi

SAY'S STINK BUG

ADULT Above, pale green with white speckling; scutellum with three distinct, white spots anteriorly and pale yellow or white tip; pronotum and abdomen edged with light yellow and white. Similar to *Chlorochroa uhleri* but latter is darker green with no white speckling and its scutellum has two white spots anteriorly. **BODY LENGTH** 15 mm. **FOOD** Adult and nymph: Plants. **FOUND** Throughout the region.

Chlorochroa sayi

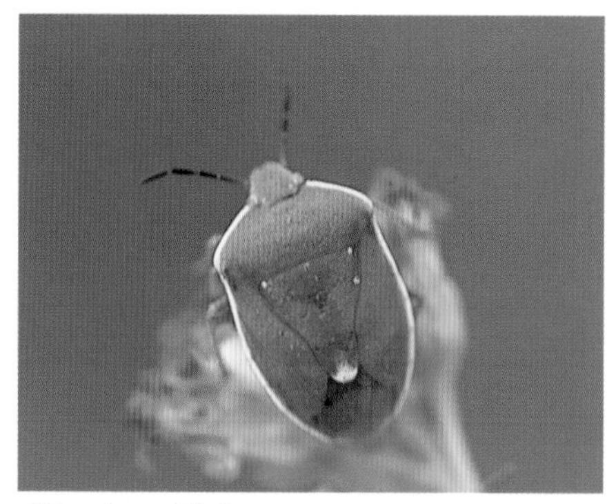

Chlorochroa uhleri

Chlorochroa uhleri, nymph

Cosmopepla conspicillaris, nymphs and adult

Chlorochroa uhleri

UHLER'S STINK BUG

ADULT Green; scutellum with white to pale yellow tip; pronotum and abdomen edged with white to pale yellow. Similar to *Chlorochroa sayi* but latter is paler green with white speckling and its scutellum has three white spots anteriorly; narrower and slightly shorter than *Acrosternum hilaris*. **BODY LENGTH** 14 mm. **NYMPH** Green with black speckling; pronotum with pale pink lateral margins; abdomen with pale pink dorsal spots and pale pink lateral margins. **FOOD** Adult and nymph: Plants. **FOUND** Common throughout the region. Look for adults on flowers in late summer and fall.

This species and *Chlorochroa ligata* are two of the most common large stink bugs in the region.

Cosmopepla conspicillaris

CONSPICUOUS STINK BUG

ADULT Black with narrow white margins; pronotum with orange band surrounding two black spots; scutellum with white-edged tip. **BODY LENGTH** 7 mm. **NYMPH** Head, pronotum, and wingpads black; abdomen light yellow with black spots. **FOOD** Adult and nymph: Plants. **FOUND** Throughout the region.

Pete has seen adults and nymphs in large numbers on hedge nettle (*Stachys* spp.) and singly on berry vines.

Elasmostethus cruciata

ADULT Above, yellowish green with reddish brown X. **BODY LENGTH** 11 mm. **NYMPH** Early instars—head, pronotum, and wingpads black; abdomen dark red with black spots; later instars—head, pronotum, and wingpads black; abdomen green with black spots and red markings. **FOOD** Adult and nymph: Red alder (*Alnus rubra*) and other alders (*Alnus* spp.) (immature cones). **FOUND** Throughout the region.

Some entomologists place this species in the family Acanthosomatidae. Adults and nymphs are often found feeding together on the same alder cone.

Elasmostethus cruciata

Elasmostethus cruciata, nymphs

Elasmucha lateralis

ADULT Mottled with dark brown and reddish brown; scutellum with white tip. **BODY LENGTH** 8 mm. **NYMPH** Later instars, head, pronotum, and wingpads black; abdomen dark red with black spots and pale yellow line on each side. **FOOD** Adult and nymph: Red alder (*Alnus rubra*) and other alders (*Alnus* spp.). **FOUND** Throughout much of the northern U.S.

The female reportedly cares for her eggs and young.

Elasmucha lateralis, nymph and adult

Euschistus conspersus

CONSPERSUS STINK BUG

ADULT Above, light brown with black speckling. **BODY LENGTH** 12 mm. **FOOD** Adult and nymph: Plants. **FOUND** Throughout much of the U.S.

This is one of the more conspicuous plant-feeding stink bugs in the region.

Euschistus conspersus

Murgantia histrionica

Murgantia histrionica, nymph

Perillus bioculatus

Murgantia histrionica

HARLEQUIN BUG

ADULT Variably patterned with black and orange. **BODY LENGTH** 11 mm. **NYMPH** Variably patterned with black and orange; abdomen with white transverse lines. **FOOD** Adult and nymph: Mustard family (Brassicaceae). **FOUND** Oregon, California, and much of the northeastern U.S.

Perillus bioculatus

TWO-SPOTTED STINK BUG

ADULT Black; pronotum and scutellum with variable patterns of white to red markings (common pattern shown). **BODY LENGTH** 9 mm. **FOOD** Adult and nymph: Invertebrates. **FOUND** Widespread throughout the U.S.

Podisus maculiventris

SPINED STINK BUG

ADULT Yellowish brown to reddish brown with black speckling. **BODY LENGTH** 12 mm. **NYMPH** Later stages, head, pronotum, and wingpads black; abdomen creamy white bordered with black and with black spots and dark red transverse lines. **FOOD** Adult and nymph: Invertebrates. **FOUND** Throughout the region.

This is the most common predaceous stink bug in the region.

Podisus maculiventris

Podisus maculiventris, nymph

Thyanta pallidovirens

WESTERN RED-SHOULDERED STINK BUG

ADULT Green; pronotum with dark red transverse band (may be much reduced); scutellum with dark red tip. **BODY LENGTH** 10 mm. **FOOD** Adult and nymph: Plants. **FOUND** Throughout the U.S.

Thyanta pallidovirens, mating pair

Zicrona caeruleus

ADULT Metallic dark blue. **BODY LENGTH** 7 mm. **FOOD** Adult and nymph: Invertebrates. **FOUND** Throughout the northern U.S. Adults common on willows in late summer and fall.

Zicrona caeruleus

AMBUSH BUGS Family Phymatidae

These small- to medium-sized, greenish yellow or brown and yellow, stout-bodied bugs are usually found hiding in flowers, well camouflaged by their shape and color. There they wait to ambush insect prey, mainly bees, wasps, and flies, that are often much larger than they are. They use their stout, raptorial front legs to grab and hold their victims. They immobilize the prey by injecting saliva through their piercing mouthparts and then proceed to suck them dry.

Phymata pacifica

PACIFIC AMBUSH BUG

ADULT Above, pale brown with white patches and brown bands. **BODY LENGTH** 12 mm. **FOOD** Adult and nymph: Invertebrates. **FOUND** Throughout the region.

Phymata pacifica

ASSASSIN BUGS Family Reduviidae

These medium- to large-sized bugs are usually black, brown, or red in color and elongate in shape. The head is narrow and elongate, with the part behind the eyes necklike. The abdomen is often widened at the middle. Some species are very long and slender, with long legs and long antennae. Many species are covered with spines or hairs and are sticky, which causes material to adhere to their bodies, thus helping to camouflage them. Most species prey on other insects, using their raptorial front legs to grab and hold their prey while they suck out its body fluids. These bugs get their common name from the way they attack and pierce their victims with their mouthparts. Some species suck blood from vertebrates; a few transmit diseases to humans. Many can inflict a painful bite if handled.

Apiomerus sp.

Apiomerus spp.

ROBUST ASSASSIN BUG

ADULT Variably patterned with red, black, and yellow (common coloration and pattern shown). **BODY LENGTH** 13 mm. **NYMPH** Variably patterned with red, black, and yellow. **FOOD** Adult and nymph: Invertebrates. **FOUND** Throughout the region, commonly on or in flowers.

This bug is fairly easy to identify to genus but difficult to identify to species.

Sinea diadema

Sinea diadema

SPINED ASSASSIN BUG

ADULT Pale brown to brown; pronotum spiny and with toothed lateral margins; head and front legs spiny. **BODY LENGTH** 14 mm. **FOOD** Adult and nymph: Invertebrates. **FOUND** Throughout the U.S.

This is one of the most common assassin bugs in the region.

Zelus renardii

LEAFHOPPER ASSASSIN BUG

ADULT Pronotum and wingcovers pale brown; abdomen and legs pale green. **BODY LENGTH** 13 mm. **FOOD** Adult and nymph: Invertebrates. **FOUND** Throughout the region. Look for adults stalking their prey on trees and shrubs.

Zelus renardii

SCENTLESS PLANT BUGS Family Rhopalidae

These bugs are very similar to squash bugs (Coreidae) but are smaller, usually paler in color, and lack scent glands. They usually occur on weeds and grasses, where the adults and nymphs feed in late summer and early fall; they can be commonly caught in a sweep net at this time of the year.

Boisea rubrolineatus

WESTERN BOX ELDER BUG

ADULT Dark brown to black; pronotum and wings with conspicuous red lines; abdomen red. **BODY LENGTH** 13 mm. **NYMPH** Head, pronotum, and wing pads black; abdomen dark red. **FOOD** Adult and nymph: Box elder (*Acer negundo*) and other maples (*Acer* spp.), ash (*Fraxinus* spp.). **FOUND** Throughout the region, often in large numbers. Overwinter as adults; frequently found sunning themselves on a rock or side of a tree on warm winter days.

Also known as *Leptocoris trivittatus*.

Boisea rubrolineatus

Boisea rubrolineatus, nymph

CICADAS, LEAFHOPPERS, AND ALLIES
Order Homoptera

Homoptera (Greek, *homo*, same; *ptera*, wings) is a large and diverse order that contains approximately 32,000 species worldwide, with over 6350 species recorded in the United States and Canada. This order is closely related to Hemiptera—so closely, in fact, that some entomologists group the two orders into one, Heteroptera. They range in size from small to large; the majority in the Pacific Northwest are small. Despite their small size, homopterans in general have a very significant impact on plants since the individuals of many species are enormously abundant. It is almost impossible to find a plant, at one time or another, that does not have a population of homopterans, such as aphids, living on it.

Although the adults in many species are winged, there are some species in which one or both sexes may be wingless. Winged homopterans usually have two pairs of wings with forewings that are more or less uniformly membranous and thin (hence the order name) and hindwings that are membranous. The wings are usually held rooflike over the body when at rest.

Most homopterans have functioning mouthparts. In those that do, the mouthparts have been modified into a beak for piercing and sucking. The beak arises from back beneath the head instead of the front of the head, as in the bugs (Hemiptera).

Homopterans exhibit considerable variation in body form. In many species, the body is soft, but in others it is thickened and leathery, sometimes with bristles or spines. Some secrete waxy substances that may cover the body or form projecting filaments or frothy masses. Many also secrete from their anus a saplike liquid called *honeydew*, which attracts other insects, such as ants and wasps. Because of the large number of species within the order and the small size of most species, this order as a whole is difficult to identify to species.

High rates of reproduction are a rule in Homoptera, although the means of reproduction varies considerably within the order. Some species reproduce sexually, whereas others do so parthenogenetically. Some species have very complex life cycles, which involve both sexual and parthenogenetic generations. Metamorphosis is usually incomplete. The immature stage, which is called a *nymph*, normally resembles a wingless adult.

All homopterans feed on plant fluids. They are rarely found far from their host plants; in fact, some species, like scales, are permanently attached to their hosts.

Because they occur in such abundant numbers and because they damage plants by not only feeding on them but also by transmitting plant diseases, this group is one of the most destructive to crops and gardens. However, by virtue of their high reproductive rate, they are a major food source for many predaceous and parasitic insects, and through their production of honeydew, they provide a high-energy alternate food source for other insects.

Key to Homoptera Families

PINE & SPRUCE ADELGIDS
Family Adelgidae, page 110

APHIDS
Family Aphididae, page 111

SPITTLEBUGS
Family Cercopidae, page 113

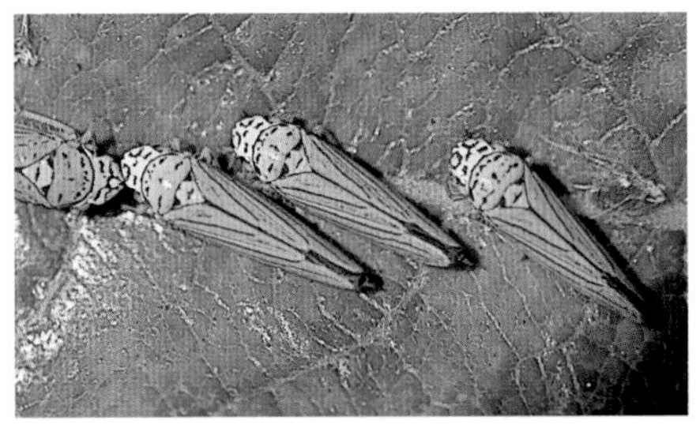

LEAFHOPPERS
Family Cicadellidae, page 114

CICADAS
Family Cicadidae, page 115

SOFT SCALES
Family Coccidae, page 116

TREEHOPPERS
Family Membracidae, page 117

PINE & SPRUCE ADELGIDS Family Adelgidae (Phylloxeridae)

This very small family consists of small, aphidlike insects. The wingless females and nymphs are often covered with a cottony, waxy coating. They are confined to conifers, feeding on twigs or needles or inside the galls they form. Most species alternate between two different species of conifers but form galls only on the primary host tree. Some species have up to six different forms of egg-laying females. Adelgids have spread worldwide as forest and ornamental plant pests.

Adelges cooleyi on Douglas-fir

Adelges cooleyi, galls on spruce

Adelges cooleyi

COOLEY SPRUCE GALL APHID

ADULT AND NYMPH (on Douglas-fir): Covered with white waxy coating. **GALL** (on spruce): Pineapple- or conifer cone–shaped; green to dark purple (adult and nymphal aphids feed inside internal chambers of gall). **GALL LENGTH** 20–70 mm. **FOOD** Adult and nymph: Sitka spruce (*Picea sitchensis*) and other spruces (*Picea* spp.) (buds), Douglas-fir (*Pseudotsuga menziesii*) (needles). **FOUND** Originally in the western U.S., now found throughout the U.S. and Europe.

The worst infestations of Douglas-fir by this species usually occur in ornamental plantings where the trees are under stress. In severe infestations the trees may resemble flocked Christmas trees. Damaged needles may fall prematurely.

Adelges cooleyi, interior of gall with aphids

Adelges tsugae

HEMLOCK WOOLLY APHID

ADULT Covered with white waxy coating. **BODY LENGTH** 2 mm. **FOOD** Adult and nymph: Western hemlock (*Tsuga heterophylla*). **FOUND** Throughout the region. They occur in small white masses that are most noticeable around the base of needles; they are harder to see on bark and twigs.

Adelges tsugae

APHIDS Family Aphididae

The members of this large, well-known family are found almost everywhere and frequently in large numbers. They are very small, usually pale in color, and have a soft, pear-shaped body with a small head and long antennae. Some adults have wings; others are wingless. Most species have a pair of slender tubes, called *cornicles,* that project from the back of the abdomen and secrete a defensive fluid when the aphid is disturbed. Aphids also produce honeydew, which attracts other insects; some ant species even go so far as to form a symbiotic relationship with the aphids—the ants tend to the aphids like "cows," moving them to suitable host plants and protecting them from predators and parasites in return for the honeydew. Aphids may produce honeydew in such quantities as to cause the surface of objects below them to become sticky.

Most species have a complex life cycle. There are two basic types of life cycle, but the following is typical of most species. Overwintering eggs hatch in the spring and develop into wingless females, which give birth parthenogenetically to young that mature into wingless females. Generations of wingless females are produced in this way for as long as environmental conditions allow. A final generation of winged females is then produced, and these females migrate to new host plants. During the fall, they produce wingless males and egg-laying females, which mate. The females lay eggs on the host plants, and the cycle begins again.

Aphids feed on the sap of plant stems, roots, leaves, and flowers. Some species form galls. Many species are major agricultural and garden pests: they reproduce so rapidly and in such large numbers that they can destroy entire crops.

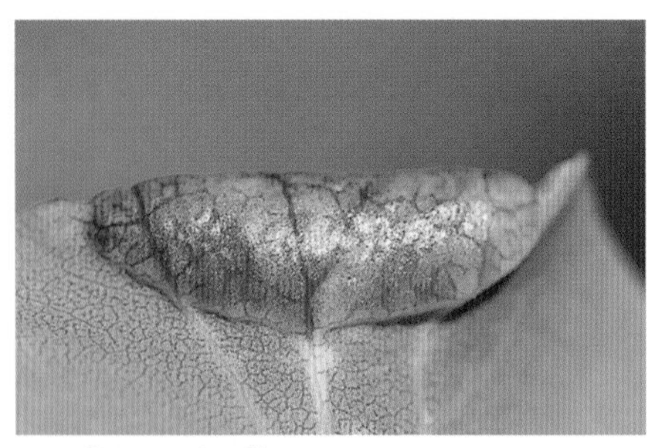

Tamalia coweni, gall

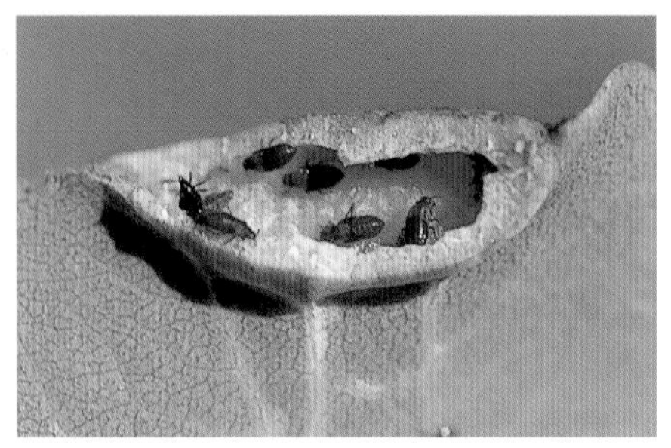

Tamalia coweni, interior of gall with aphids

Thecabius populimonilis, galls

Tuberolachnus salignus

Tamalia coweni

MANZANITA LEAF GALL

ADULT AND NYMPH Black. **GALL** Red; formed from folded edge of leaf; several aphids per gall. **GALL LENGTH** 10 mm. **FOOD** Adult and nymph: Manzanita (*Arctostaphylos* spp.). **FOUND** Range unknown but probably same as host.

Thecabius populimonilis

POCKET LEAF GALL

GALL Red; oriented parallel to midvein of leaf and with opening on underside of leaf. **GALL LENGTH** 5–15 mm. **FOOD** Adult and nymph: Black cottonwood (*Populus balsamifera*). **FOUND** Throughout the region.

Also known as *Pemphigus populivnae*. Spring galls are small and few in number per leaf; summer galls are larger and often arranged end to end down the length of the leaf, which may be greatly distorted.

Tuberolachnus salignus

GIANT WILLOW APHID

ADULT Grayish brown with black spots and covered with fine, waxy coating; abdomen with large, black central protuberance. **BODY LENGTH** 4 mm. **FOOD** Adult: Coastal willow (*Salix hookeriana*), shining willow (*S. lucida*), and other willows (*Salix* spp.). **FOUND** Throughout the U.S.

Also known as *Pterochlorus viminalis*. This species is one of the largest aphids on the West Coast. It feeds in colonies on the trunk of the host plant. When a colony is threatened, all the individuals of the colony energetically kick their back legs. This action probably helps to scare off potential predators and parasites.

SPITTLEBUGS Family Cercopidae

The members of this large family are very similar to leafhoppers (Cicadellidae) but are more robust and usually less colorful. They are small, usually brown or gray, with forewings that are longer than the body and often very thick and leathery. The hind legs are usually enlarged and elongate and adapted for leaping. The adults are often called *froghoppers* because they vaguely resemble tiny frogs. The females lay their eggs in the stems or sheaths of grasses and other plants. The nymphs are often most abundant in late spring, but what one is more likely to see than the nymphs themselves, or the adults, is the frothy, spittlelike masses that the nymphs produce (and which account for the common name for the family). The spittle is secreted from the anus and mixes with a substance excreted from the abdomen. Air bubbles are then incorporated into the spittle. The nymph usually rests head downward on a plant, and as the spittle forms, it flows down over and completely covers the body. Each bubbly mass contains one or more nymphs, protecting them from predators and parasites and providing a moist habitat. The adults do not produce spittle. Most species suck the sap from herbaceous plants, but some feed on trees. Some species do considerable damage to crops.

Philaenus spumaria

MEADOW SPITTLEBUG

ADULT Color pattern extremely variable—from plain black to pale brown with white spots. **BODY LENGTH** 6 mm. **NYMPH** Head, pronotum, and wing pads black; abdomen pale yellow. **FOOD** Adult and nymph: Plants. **FOUND** Throughout the U.S.

Philaenus spumaria, nymph and spittle

■ LEAFHOPPERS Family Cicadellidae

This is the largest family of Homoptera; not only is it very abundant in species but also in individuals. They are similar in appearance to spittlebugs (Cercopidae) but have more elongate bodies and are usually more colorful. They vary in shape and size; most are small. They also vary in color but most often are green or brown, usually with brightly colored markings. The hind legs are adapted for leaping. The adults have forewings that are thick and leathery. Leafhoppers excrete honeydew. Both the adults and nymphs feed mainly on the leaves of almost all types of plants. Some are serious pests of many kinds of cultivated plants, causing wilted, discolored leaves and stunted growth. Some species transmit plant diseases.

Graphocephala atropunctata

Graphocephala atropunctata

BLUE-GREEN LEAFHOPPER

ADULT Green or blue-green; head and pronotum with black spots or lines; scutellum yellow with black markings; wings with black lines. **BODY LENGTH** 6 mm. **FOOD** Adult and nymph: Plants. **FOUND** Throughout the region.

Also known as *Hordnia circellata*. A heavy infestation of the host plant by this species will cause its leaves to turn pale. This leafhopper produces copious amounts of honeydew.

CICADAS Family Cicadidae

Members of this family can usually be recognized by their characteristic shape and large size; this family contains the largest homopteran in the United States. The body is green or brown, often with black markings. The large wings are transparent brown, black, or green, with distinct black veins, and held tentlike over the body; the forewings are twice the length of the hindwings. This family is also known for the loud buzzing or pulsating, clacking sound produced by some species that is characteristic of hot summer days and warm summer evenings. The males produce this sound, and each species has its own distinctive song. All species have a long life cycle of four or more years, with most of the life cycle spent as a nymph. The female uses her sharp ovipositor to make punctures in twigs, into which she deposits her eggs. After the nymphs emerge from the eggs, they drop to the ground, enter the soil, and feed on the tree roots. During the last instar, they dig their way out of the ground and climb a tree, where they emerge as adults.

Platypedia minor

MINOR CICADA

ADULT Bronzy black. **BODY LENGTH** 19 mm. **FOOD** Adult: None. Nymph: Trees and shrubs (roots). **FOUND** Oregon and California, most commonly in open woodlands.

Platypedia minor

SOFT SCALES Family Coccidae

This is a very large family of small, plant-feeding insects. The females are the ones usually seen—they are flattened, elongate-oval with a hard, smooth exoskeleton or covered with wax. In many species, the gravid females are convex, somewhat like an abalone shell. The females of some species are legless. Unlike the females, the males look like "typical" insects, although in some species they are wingless. Many species are pests of agricultural and ornamental trees and shrubs.

Parthenolecanium corni, females and nymphs

Physokermes concolor, female

Physokermes hemicryphus, female

Parthenolecanium corni

EUROPEAN FRUIT SCALE

ADULT Female, convex in shape; coloration variable, mature female reddish brown. **BODY LENGTH** 7 mm. **NYMPH** Flattened convex in shape; coloration variable. **FOOD** Adult and nymph: Plants. **FOUND** Throughout the region.

Although this is an introduced species, it is so widespread throughout the U.S., it warrants inclusion. Its infestations may occasionally cause damage or death to host plants. The males of this species are not common; the females reproduce parthenogenetically.

Physokermes concolor

FIR BUD SCALE

ADULT Female, oval, budlike; shiny reddish brown when fresh. **BODY LENGTH** 5 mm. **FOOD** Adult and nymph: Grand fir (*Abies grandis*), white fir (*A. concolor*), and other firs (*Abies* spp.). **FOUND** Throughout the region.

The females are easiest to find in late fall to early spring. When alive, the scale is often attended by ants.

Physokermes hemicryphus

SPRUCE BUD SCALE

ADULT Female, masquerades as a spruce bud—oval, budlike; shiny, pale brown with black markings when fresh. **BODY LENGTH** 8 mm. **FOOD** Adult and nymph: Sitka spruce (*Picea sitchensis*) and other spruces (*Picea* spp.). **FOUND** Throughout the region.

This species was introduced from Europe.

TREEHOPPERS Family Membracidae

The members of this large family of small, jumping insects can be recognized by the large pronotum that covers the head and extends over the abdomen, concealing the wings. The pronotum is often modified into a horn, spine, or keel. Many species appear humpbacked. The adults and nymphs are commonly found in mixed-age groups on trees, shrubs, and field vegetation. Most species are host specific. Some feed on grass and herbaceous plants in the nymphal stage. Treehoppers excrete copious amounts of honeydew.

Platycotis vittatus

OAK TREEHOPPER

ADULT Male, pronotum mostly white with reddish orange lines and elongated anteriorly into horn; wings translucent with black veins and orange along leading margin; eyes red. Female, dark brown; similar to male in shape. **BODY LENGTH** 8 mm. **NYMPH** Pronotum dark brown with pair of dorsal spines; wing pads dark brown; abdomen black with white-and-red transverse lines; eyes red. **FOOD** Adult and nymph: Red alder (*Alnus rubra*) and other alders (*Alnus* spp.), oak (*Quercus* spp.), and many other broadleaf trees. **FOUND** Throughout the region. To find a colony, look directly above wet patches of honeydew on leaves or ground.

Platycotis vittatus, male

Platycotis vittatus, female and nymphs

WASPS, ANTS, AND BEES Order Hymenoptera

Hymenoptera (Greek, *hymen*, membrane; *ptera*, wings) is the third-largest order of insects, with approximately 103,000 species described worldwide and, of those, over 17,700 known in the United States and Canada. Hymenopterans exhibit great diversity in habit and complexity of behavior, with some having highly developed social systems. From a human standpoint, this order is probably considered the most beneficial of all the insect orders.

The adults range in size from tiny (egg parasites) to large. Most adults are winged; they have two pairs of membranous wings that have relatively few veins (hence the order name), with the forewings larger than the hindwings. In some species, the adults are wingless in one or both sexes or in certain castes. Many, but not all, have a constriction, or "waist," between the thorax and abdomen. Hymenopterans usually have chewing mouthparts that in some species are also modified into tonguelike structures for lapping or sucking plant liquids. The females of many species have a well-developed ovipositor. In most bees and predatory wasps and in some ants, the ovipositor is modified into a stinger.

Hymenopterans undergo complete metamorphosis. The larvae occur in a great variety of forms. In most species, the larvae are legless, appearing grublike or maggotlike; in some, they are caterpillarlike. They all have a well-developed head with chewing mouthparts.

The social behavior of hymenopterans ranges from solitary lifestyles, in the majority of species, to the most complex social organizations in the insect class. The physical structure of the nests or colonies and the caste systems within the nests are extremely varied and highly developed in several families, particularly the ants, social wasps, and social bees. The caste systems usually include a fertile queen(s), fertile males, and sterile females that are workers, and, in some cases, soldiers. The workers, which constitute the largest caste, perform such functions as building, maintaining and defending the nest, gathering the food, and tending to the queen and the nursery (eggs, larvae, and pupae).

In many species, the adults feed on pollen and nectar; many others prey on other insects, mainly for the purpose of providing food for the developing larvae. The larvae of a great number of species are internal or external parasites of other insects. The larvae of others feed on plants; those of bees and a few wasps feed on pollen and nectar.

Hymenopterans play various significant roles in ecological processes. The adults of bee and many wasp species may be the most important insects that are involved in the pollination of plants and consequently in maintaining the diversity of plant life and increasing the production of crops. In fact, many plants completely depend on bees for pollination, and hence for survival. Hymenopterans are also essential in controlling insect populations, including those of crop pests, through predation and parasitism. One family of hymenopterans, the ants, plays a vital role in the biological and physical health of soils. Another important economic benefit to humans that bees provide is their production of honey and wax.

Hymenopterans are less important than other orders as economic pests. The larvae of some species damage trees and crops by boring into the plant or by feeding on the leaves or conifer needles.

Key to Hymenoptera Families

BUMBLE BEES & HONEY BEES
Family Apidae, page 120

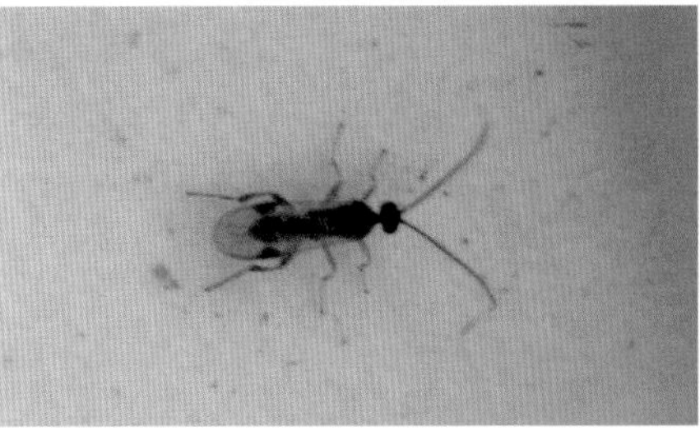

BRACONIDS
Family Braconidae, page 121

CIMBICID SAWFLIES
Family Cimbicidae, page 122

GALL WASPS
Family Cynipidae, page 123

ANTS
Family Formicidae, page 129

ICHNEUMONFLIES
Family Ichneumonidae, page 130

HORNTAILS
Family Siricidae, page 131

DIGGER WASPS
Family Sphecidae, page 132

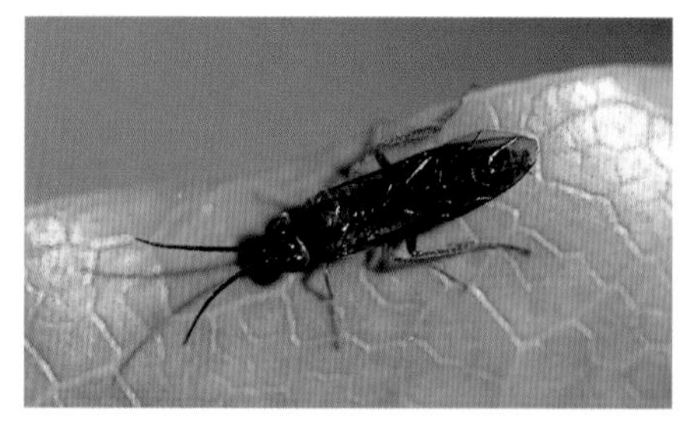

COMMON SAWFLIES
Family Tenthredinidae, page 133

TRUE WASPS
Family Vespidae, page 135

BUMBLE BEES & HONEY BEES Family Apidae

The members of this well-known family of social insects are black, often with yellow markings, or brown, and have a yellow or yellow and black pubescence (sometimes it is partly reddish brown to nearly orange). The females have an ovipositor that is modified into a stinger. Although they have a wide range of habits, including parasitism, both bumble bees and honey bees live in colonies consisting of one queen, female workers, and males, which are often called *drones*. Most species produce and store honey. Although honey bees, which are not native to North America, are essential pollinators of commercial crops, many native plants have special pollination needs that can be met only by bumble bees, which are native. The loss or diminution of native bee populations can have a dramatic negative effect on native plant populations.

Bombus mixtus

Bombus mixtus

MIXED BUMBLE BEE

ADULT Head and thorax covered with yellow and black hairs; abdomen banded with yellow and black hairs and with reddish brown tip. **BODY LENGTH** 13 mm. **FOOD** Adult: Plants (nectar). **FOUND** From British Columbia to California.

Bombus vosnesenskii

YELLOW-FACED BUMBLE BEE

ADULT Covered with black hair except for face and shoulders, which are covered with yellow hair, and narrow yellow band of hairs near posterior end of abdomen. **BODY LENGTH** 22 mm. **FOOD** Adult: Plants (nectar). **FOUND** Throughout the region.

This species is one of the most common bumble bees.

Bombus vosnesenskii

BRACONIDS Family Braconidae

The members of this large family are small parasitic wasps. They are similar to ichneumonflies (Ichneumonidae) but are usually smaller and stockier and have different wing venation. They are usually black or brown in color. The females lay their eggs on other species of insects, especially the caterpillars of many Lepidoptera species. After hatching from the eggs, the larvae enter the host and consume it. When they mature, they burrow out of the host and pupate in silken cocoons attached to the host's body. Some species are used to control agricultural pests.

Dinocampus coccinellae

LADYBIRD BEETLE WASP

ADULT Black. **BODY LENGTH** 4 mm. **LARVA** Tiny; white. **PUPA** Pale brown within lacy white cocoon. **FOOD** Larva: Ladybird beetles (parasitic). **FOUND** Throughout the region.

Also known as *Perlitus coccinellae*. Look for adult ladybird beetles that are lethargic or show no movement, even when prodded, but are still alive. With patience you may see the parasitic wasp larva emerge from the beetle. Since the larva does not consume the beetle's vital organs, the beetle usually lives for a short period of time after the larva emerges. Once outside the beetle, the larva spins a cocoon under the beetle and pupates.

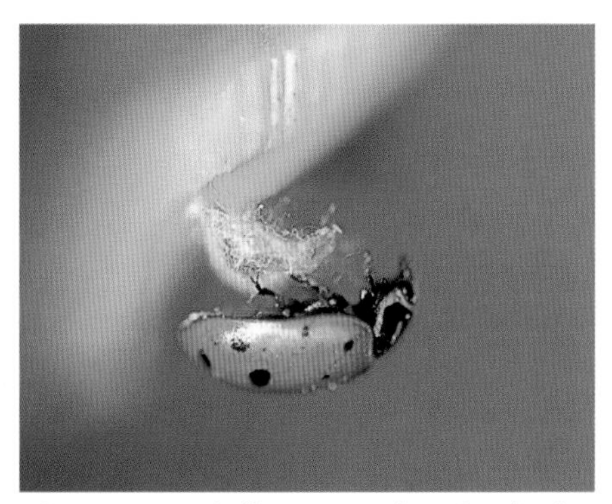

Dinocampus coccinellae, pupa under *Hippodamia convergens*

CIMBICID SAWFLIES Family Cimbicidae

These sawflies are large, robust, and bumble bee–like, and have clubbed antennae. As in other sawfly families, the thorax and abdomen are broadly joined and not separated by a slender waist. The adults feed on nectar and water, although some may damage the tips of trees. The larvae feed on the foliage of trees and pupate in silken cocoons, either in the ground or attached to twigs.

Cimbex americana, adult atop cocoon

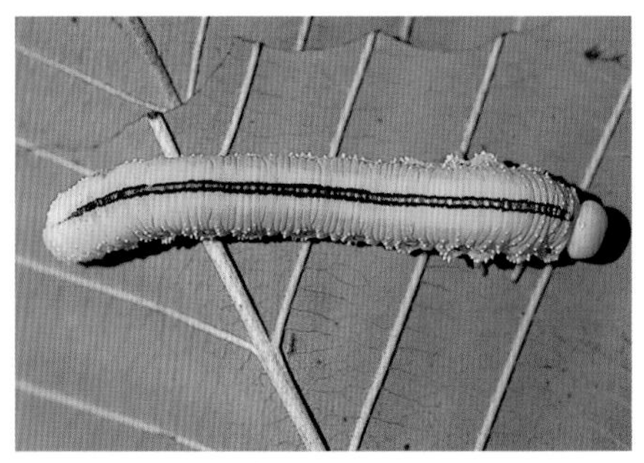
Cimbex americana, larva

Cimbex americana

AMERICAN ELM SAWFLY

ADULT Black; abdomen with yellow spots and yellow dorsal band. **BODY LENGTH** 25 mm. **LARVA** Yellowish green with blue longitudinal dorsal line edged in black. Similar to *Trichiosoma triangulum*; latter lacks blue longitudinal dorsal line edged in black. **FOOD** Larva: Coastal willow (*Salix hookeriana*) and other willows (*Salix* spp.), red alder (*Alnus rubra*) and other alders (*Alnus* spp.). **FOUND** Throughout the U.S. Adult is hard to find, but larva is common, often resting in a tight coil, holding on to underside of leaf with its front legs.

The *Cimbex americana* that Pete successfully raised remained dormant as a larva in a cocoon case for two years before pupating and emerging.

Trichiosoma triangulum

ADULT Black, covered with light yellow hairs; antennae clubbed. **BODY LENGTH** 20 mm.

Trichiosoma triangulum

Trichiosoma triangulum, larva

LARVA Green with granular skin; head pale yellow with large brown spots. Similar to *Cimbex americana*; latter has blue longitudinal dorsal line edged in black. **FOOD** Larva: Willow (*Salix* spp.). **FOUND** Throughout the region.

GALL WASPS Family Cynipidae

The members of this large, common family are tiny wasps that are usually humpbacked and black or brown in color. The abdomen is characteristically oval and somewhat flattened laterally. Most species form galls; others inhabit galls as inquilines in the larval stage. The females of most species lay their eggs in a specific host plant. The plant then forms a swelling, or gall, within which the larvae feed on the internal plant tissues. The larvae have a blind gut and excrete no fecal matter until pupation. The galls are distinctive in appearance and are the best means for identifying species.

Andricus atrimentus

DUNCE CAP GALL

GALL Cone-shaped; white striped with red. **GALL HEIGHT** 4 mm. **FOOD** Larva: Blue oak (*Quercus douglasii*) and other oaks (*Quercus* spp.). **FOUND** Range unknown but probably same as host.

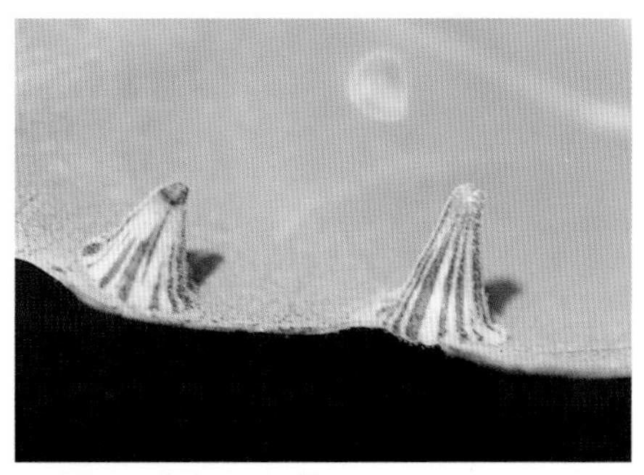

Andricus atrimentus, galls

Andricus brunneus

GALL Round, brown, and pubescent; interior of fresh gall purplish; attached to midvein on underside of leaf. **GALL WIDTH** 6 mm. **FOOD** Larva: Oregon oak (*Quercus garryana*) and other oaks (*Quercus* spp.). **FOUND** Range unknown but probably same as host.

The adult emerges in the fall.

Andricus brunneus, galls

Andricus crystallinus, galls

Andricus crystallinus

CRYSTALLINE GALL

GALL Reddish brown or white hairy mass with crystallinelike rods showing above the hairs. **GALL HEIGHT** 14 mm. **FOOD** Larva: Oregon oak (*Quercus garryana*) and other oaks (*Quercus* spp.). **FOUND** Range unknown but probably same as host.

Andricus kingi, gall

Andricus kingi

RED CONE GALL

GALL Cone-shaped; red. **GALL HEIGHT** 5 mm. **FOOD** Larva: Oregon oak (*Quercus garryana*) and other oaks (*Quercus* spp.). **FOUND** Range unknown but probably same as host.

Andricus parmula, galls

Andricus parmula

DISC GALL

GALL Disc-shaped with short stalk; reddish brown; attached to leaf. **GALL WIDTH** 3 mm. **FOOD** Larva: Oregon oak (*Quercus garryana*) and other oaks (*Quercus* spp.). **FOUND** Range unknown but probably same as host.

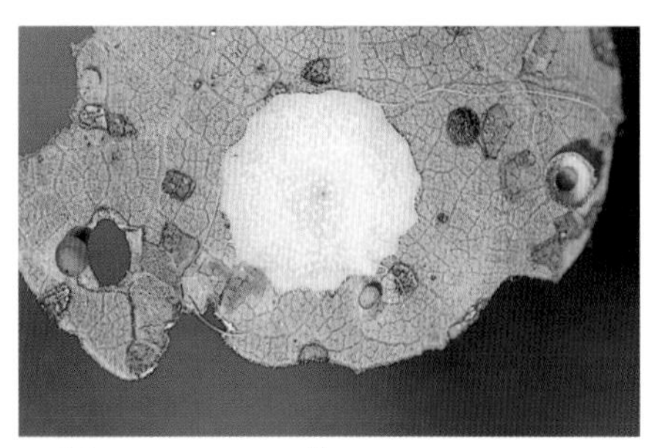
Andricus pattersonae, gall

Andricus pattersonae

GALL Flat, thin; yellow; margins smooth or toothed; attached to underside of leaf. **GALL WIDTH** 9 mm. **FOOD** Larva: Oregon oak (*Quercus garryana*) and other oaks (*Quercus* spp.). **FOUND** Range unknown but probably same as host.

Andricus stellaris

SUNBURST GALL

GALL White crystallinelike projections radiating from pinkish red to brown central disk; attached to underside of leaf. **GALL WIDTH** 4 mm. **FOOD** Larva: Oregon oak (*Quercus garryana*) and other oaks (*Quercus* spp.). **FOUND** Range unknown but probably same as host.

Andricus stellaris, galls

Antron douglasii

STAR GALL

GALL Star-shaped with blunt-tipped spines and distinct stalk; light pink to dark red; attached to leaf. Similar to *Antron quercusechinus*; latter lacks stalk. **GALL WIDTH** 15 mm. **FOOD** Larva: Blue oak (*Quercus douglasii*) and other oaks (*Quercus* spp.). **FOUND** Range unknown but probably same as host.

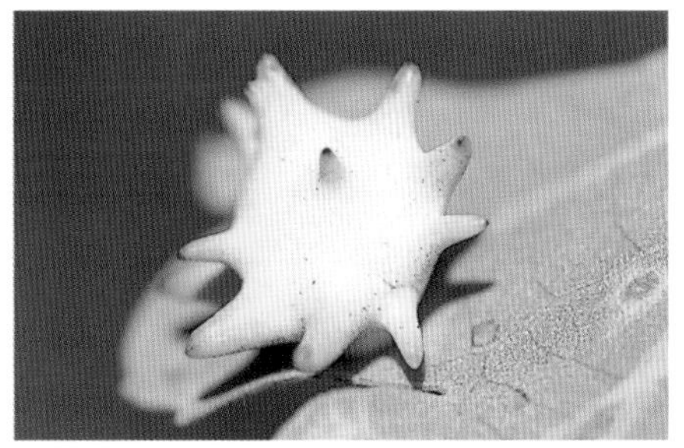

Antron douglasii, gall

Antron quercusechinus

ECHINID OAK GALL

GALL Covered with blunt spines, the tips of which are often slightly bent; pink to dark red. Similar to *Antron douglasii*; latter has distinct stalk. **GALL WIDTH** 10 mm. **FOOD** Larva: Blue oak (*Quercus douglasii*) and other oaks (*Quercus* spp.). **FOUND** Range unknown but probably same as host.

Also known as *Antron echinus*.

Antron quercusechinus, galls

Besbicus mirabilis

GALL Round; green or pale yellow with reddish brown spots; attached to leaf; larva suspended in center by radiating fibers connected to thin outer wall (see *Trichoteras vacciniifoliae*). **GALL WIDTH** 30 mm. **FOOD** Larva: Oregon oak (*Quercus garryana*). **FOUND** Range unknown but probably same as host.

Besbicus mirabilis, gall

Diastrophus kincaidii, previous year's gall and fresh (green) gall

Diplolepis polita, galls

Diplolepis rosae, galls

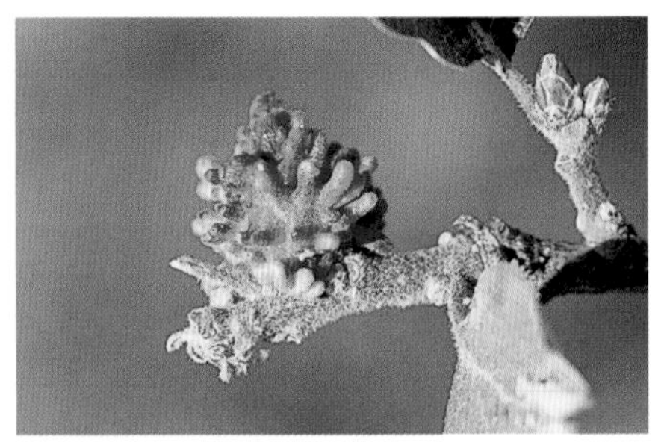

Disholcaspis corallina, gall

Diastrophus kincaidii

THIMBLEBERRY STEM GALL

GALL Enlarged along stem; green when new, brown and woody as matures; in second year of growth, emergence holes of adult wasps apparent (see photograph). **GALL LENGTH** 80 mm. **FOOD** Larva: Thimbleberry (*Rubus parviflorus*). **FOUND** Throughout the region.

In early spring, when the adults emerge from a gall, they rarely fly but instead often stay on the same plant from which they emerged, walking along the stems looking for mates.

Diplolepis polita

ROSE LEAF GALL

GALL Round; red when in sun, green when in shade; covered with long, dark red spines; attached to leaf. **GALL WIDTH** 10 mm. **FOOD** Larva: Wood rose (*Rosa gymnocarpa*) and other native roses (*Rosa* spp.). **FOUND** Oregon and California.

There may be one to many galls on one leaf.

Diplolepis rosae

MOSSY ROSE GALL

GALL Irregularly shaped and covered with mass of feathery strands; when fresh, "body" of gall red, strands greenish yellow; contains many larvae. **GALL LENGTH** 70 mm. **FOOD** Larva: Sweet-brier (*Rosa eglanteria*). **FOUND** Range unknown but probably same as host.

Disholcaspis corallina

CORAL GALL

GALL Round and covered with clublike protuberances; red, yellow, or orange when fresh. **GALL WIDTH** 10 mm. **FOOD** Larva: Blue oak (*Quercus douglasii*). **FOUND** Oregon and California.

Disholcaspis eldoradensis

HONEYDEW GALL

GALL Flat-topped oval; red with dark brown top when new, completely dark brown when mature; strongly attached to stem. **GALL WIDTH** 8 mm. **FOOD** Larva: Oregon oak (*Quercus garryana*) and other oaks (*Quercus* spp.). **FOUND** From Washington to California.

The gall secretes a sweet liquid that attracts ants, bees, and some species of butterflies (such as *Habrodais grunus*).

Disholcaspis eldoradensis, galls

Disholcaspis prehensa

CLASPING TWIG GALL

GALL Mushroom-shaped with pitted cap; brown; clasping base wraps around stem of host. **GALL HEIGHT** 10 mm. **FOOD** Larva: Oak (*Quercus* spp.). **FOUND** Range unknown but probably same as host.

The young gall exudes a liquid that attracts insects, particularly ants and wasps.

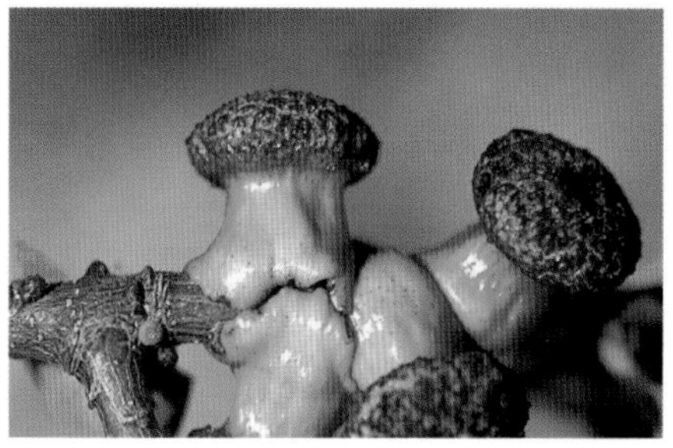

Disholcaspis prehensa, galls

Disholcaspis washingtonensis

GALL Round; pubescent; dark green when immature, pinkish red to brownish red when mature; interior fleshy with one larva in center; on stems. **GALL WIDTH** 8 mm. **FOOD** Larva: Oregon oak (*Quercus garryana*) and other oaks (*Quercus* spp.). **FOUND** Throughout the region.

Disholcaspis washingtonensis, gall

Disholcaspis washingtonensis, interior of gall with larva

Dros pedicellatus, gall

Dryocosmus castanopsidis, gall

Heteroecus pacificus, gall

Neuroterus saltatorius, galls

Dros pedicellatus

GALL Long, thin, enlarged near tip; yellowish green; attached to leaf. **GALL LENGTH** 20 mm. **FOOD** Larva: Oregon oak (*Quercus garryana*) and other oaks (*Quercus* spp.). **FOUND** Range unknown but probably same as host.

Dryocosmus castanopsidis

CHINQUAPIN GALL

GALL Round, golden yellow to brown, and pubescent; attached to male flower (catkin). **GALL WIDTH** 13 mm. **FOOD** Larva: Giant chinquapin (*Chrysolepis chrysophylla*), bush chinquapin (*C. sempervirens*). **FOUND** Range unknown but probably same as host.

Heteroecus pacificus

BEAKED SPINDLE GALL

GALL Spindle-shaped; green or red when first formed, turning brown with age; attached to end of stem; contains one larva. **GALL LENGTH** 30 mm. **FOOD** Larva: Canyon live oak (*Quercus chrysolepis*), huckleberry oak (*Q. vaccinifolia*). **FOUND** Range unknown but probably same as host.

Neuroterus saltatorius

JUMPING OAK GALL

GALL Resembles an insect egg—round, brown; attached to underside of leaf. **GALL WIDTH** 1 mm. **FOOD** Larva: Oak (*Quercus* spp.). **FOUND** Throughout much of the U.S.

After the gall drops to the ground in the fall, the larva moves around inside the gall, causing it to move around ("jump"), until the larva finds a protected spot in which to overwinter.

Trichoteras vacciniifoliae

PAPER GALL

GALL Round; light brown with brownish red reticulations; attached to stem; larva suspended in center by radiating fibers connected to thin outer wall. **GALL WIDTH** 30 mm. **FOOD** Larva: Canyon live oak (*Quercus chrysolepis*) and other oaks (*Quercus* spp.). **FOUND** Range unknown but probably same as host.

The gall forms in the spring, and by the fall is brown, wrinkled, and dry.

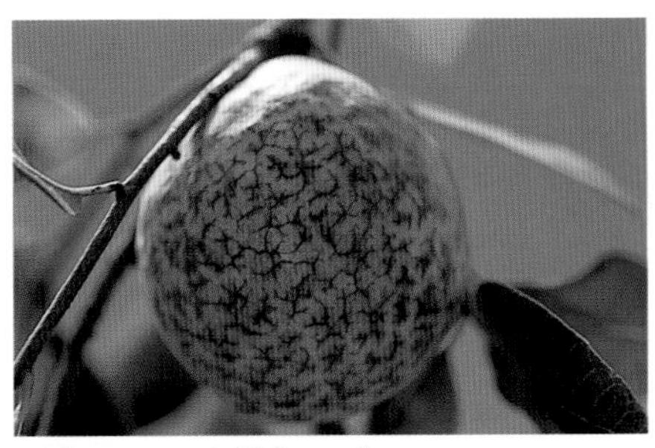

Trichoteras vacciniifoliae, gall

Trichoteras vacciniifoliae, interior of gall with larva

ANTS Family Formicidae

This well-known family is probably the most successful of all insect families, at least in temperate regions. Ants occur almost everywhere in terrestrial habitats and outnumber in individuals most other terrestrial animals. As a group they are easy to identify—they have three distinct body regions, a slender waist, and are usually black, brown, or red. They differ from wasps in having distinctly elbowed antennae. Identification to species, however, is difficult without the help of technical literature.

All ants are social insects and live in colonies that vary greatly in size. They build their nests in a great variety of places, but probably the majority of ants nest in the ground. The social structure of the colonies is complex, with most colonies composed of three castes—queen, male, and worker. The queens, which number one or more per colony, are larger than the members of the other castes and are usually winged after emerging from the pupal case. The queen(s) usually starts the colony and does most of the egg-laying in the colony (consequently all the individuals within the colony are closely related). The short-lived males are winged and usually considerably smaller than the queens. The sterile females are wingless. At certain times of the year, the queens and males swarm in a brief mating flight. After mating, the males die, and the queens shed their wings and return to the ground to start new colonies.

When threatened, a majority of species are capable of stinging or biting to defend themselves. Most species are predators or scavengers; a few harvest seeds, tend to aphids for their honeydew, raise fungus in underground gardens, or eat leaves cut from plants. Ants have a profound effect on their environment: by constructing underground colonies, they aerate the soil, and by bringing in plant and animal food items into the colonies and contributing their own waste products, they move nutrients from above ground to below ground, enriching the soil in the process.

Formica obscuripes, ants attacking intruder ant

Formica obscuripes

WESTERN THATCHING ANT

ADULT Head and thorax orange; abdomen black. **BODY LENGTH** 7 mm. **LARVA** Maggot-like; white. **FOOD** Adult: Plants and animals. **FOUND** Throughout the region.

The adults construct a nest of dried plant material that is often 28–43 cm high. They are very aggressive in procuring food and when defending the nest or the aphids they are attending. They scavenge for dead animals and prey on any live animals, usually invertebrates, that are small enough for them to overpower.

ICHNEUMONFLIES Family Ichneumonidae

These parasitic wasps are the largest family of Hymenoptera and one of the largest families of insects. The adults vary greatly in shape, size, and color, but most resemble slender wasps. They are similar to the braconid wasps (Braconidae) but are usually larger and more slender and have different wing venation. In most species, the females have an ovipositor that is long, sometimes considerably longer than the body, but they do not sting. Identification to species is usually difficult. Ichneumonflies are found nearly everywhere. The adults drink water and nectar. The larvae parasitize other insects and spiders; in some species, they are hyperparasites, that is, they parasitize other parasites, including other ichneumonids. They are important in the maintenance of insect populations.

Megarhyssa nortoni

WESTERN GIANT ICHNEUMON

ADULT Female, black; abdomen with yellow and red spots; long (50–76 mm) ovipositor. Male, less colorful than female; no ovipositor. **BODY LENGTH** 35 mm. **FOOD** Larva: Wood-boring wasp larvae (parasitic). **FOUND** Throughout the region at higher elevations. Adults may be common two or more years after a forest fire. Look for them nectaring on plants like *Angelica* spp. in mountain meadows.

The female uses her ovipositor to bore through wood and place her eggs in the subsurface tunnels of the host wasp.

Megarhyssa nortoni, female

Megarhyssa nortoni, male

HORNTAILS Family Siricidae

Horntails are large wasps with brown or black, cylindrical bodies with the thorax and abdomen broadly joined together. They are called horntails because the adults have a triangular or spearlike, horny plate at the tip of the abdomen. The females also have at the tip of their abdomen a long, slender, blunt ovipositor, which is used for drilling into wood where the eggs are then laid. After emerging from the eggs, the maggotlike larvae tunnel through the wood where they pupate. The newly formed adults emerge from the wood, sometimes stored firewood, through conspicuous holes. The adults feed on nectar and water. Horntails are not capable of stinging.

Urocerus albicornis

ADULT Black to dark blue; wings smoky black; head with white cheeks; antennae white with black base and tip; legs banded with white. **BODY LENGTH** 50 mm. **FOOD** Larva: Wood (borer). **FOUND** Throughout the region.

Urocerus albicornis

Urocerus californicus, female

Urocerus californicus

CALIFORNIA HORNTAIL

ADULT Female, black; wings reddish orange; head with yellow cheeks; antennae yellow; legs banded with yellow. Male, reddish brown. **BODY LENGTH** 50 mm. **FOOD** Larva: Wood (borer). **FOUND** Throughout the region.

DIGGER WASPS Family Sphecidae

This is a large, diverse family of solitary wasps. Most species are medium to large in size and thread-waisted. They are usually black or brown and may be patterned with other colors. Unlike true wasps (Vespidae), they have unpleated wings that do not fold over the abdomen when at rest. The adults are commonly found on flowers. They prey mostly on insects and spiders; some also feed on nectar or honeydew. Most of the predaceous species specialize in one group of arthropods (for example, lepidoptera larvae) with which to provision their nests. Although they are known as *digger wasps*, they exhibit an amazing range of nesting behavior: many do nest in underground burrows, but others nest on the ground or in above-ground habitats, such as rotten wood, hollowed out twigs, or abandoned beetle larvae borings; some construct cells of mud; a few live in the nests of other bees or wasps. All species can impart a painful sting.

Bembix americana

Bembix americana

WESTERN SAND WASP

ADULT Black; thorax with pale white hairs; abdomen with greenish white transverse bands. **BODY LENGTH** 17 mm. **FOOD** Adult: Plants (nectar). Larva: Flies. **FOUND** Throughout the region, very commonly in sandy areas.

Also known as *Bembix comata*. Although this species is not a social insect, it often aggregates in colonies. It is common to encounter a female on the sand digging a burrow or entering a completed burrow with a fly in tow. Each burrow, which leads to a nest, is 16–20 inches long, and each nest contains a larva to which the female brings fresh flies until the larva is mature. This wasp is parasitized by the fly *Physocephala texana*; it is not unusual to find *P. texana* pupae inside the dead pupae of western sand wasps.

COMMON SAWFLIES Family Tenthredinidae

This is a very large family of wasps; it is the largest family of sawflies, and its members are the most commonly found among sawflies. The "wasplike" adults are small to medium in size and black or brown, often with bright colors. The thorax and abdomen are broadly joined together, and the antennae are long and threadlike. The female's ovipositor is somewhat sawlike, hence the common name. The adults are usually found on foliage or flowers. The larvae of most species resemble those of Lepidoptera but have more prolegs. Most feed on the leaves of plants, mainly trees and shrubs. A few are either leaf miners, stem borers, or gall makers.

Hemichroa crocea

STRIPED ALDER SAWFLY

LARVA Yellow above, light green below, with black lateral lines; head black. **LARVA LENGTH** 20 mm. **FOOD** Larva: Red alder (*Alnus rubra*) and other alders (*Alnus* spp.). **FOUND** Throughout the region.

This species is introduced. The larvae periodically build up to very high numbers and strip alders of all their leaves. When a group of feeding larvae is threatened, the larvae hold onto the leaves with their true legs, wave their rear ends in the air, and, from their anal glands, exude drops of brown liquid used to repel predators and parasites.

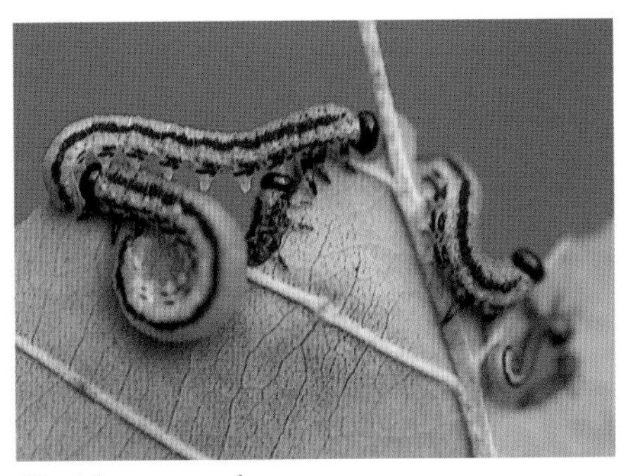

Hemichroa crocea, larvae

Nematus chalceus

ADULT Shiny reddish brown; thorax with three black spots. **BODY LENGTH** 10 mm. **LARVA** Pale green with black spots. **FOOD** Larva: Coastal willow (*Salix hookeriana*) and other willows (*Salix* spp.). **FOUND** Throughout the region.

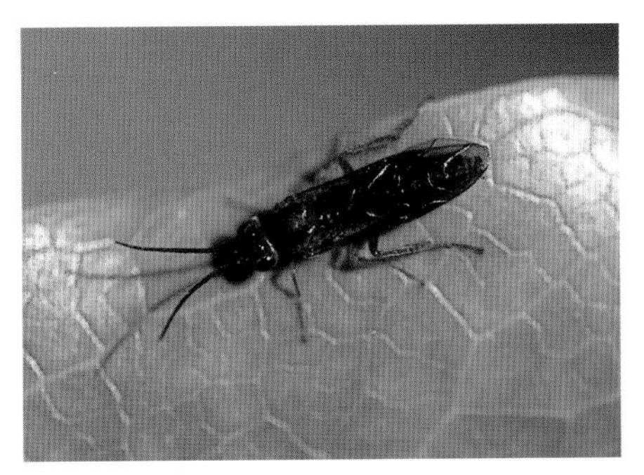

Nematus chalceus

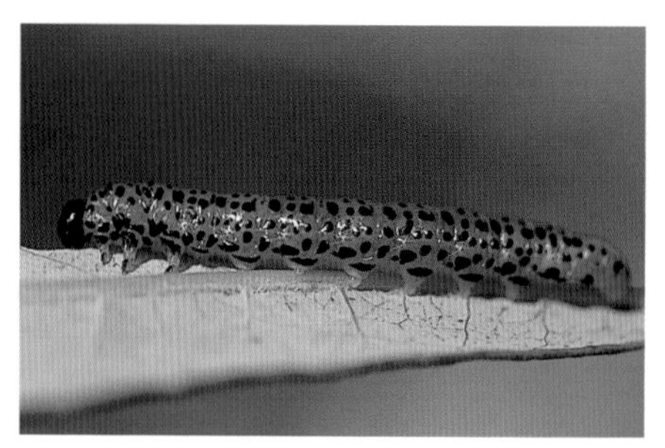

Nematus chalceus, larva

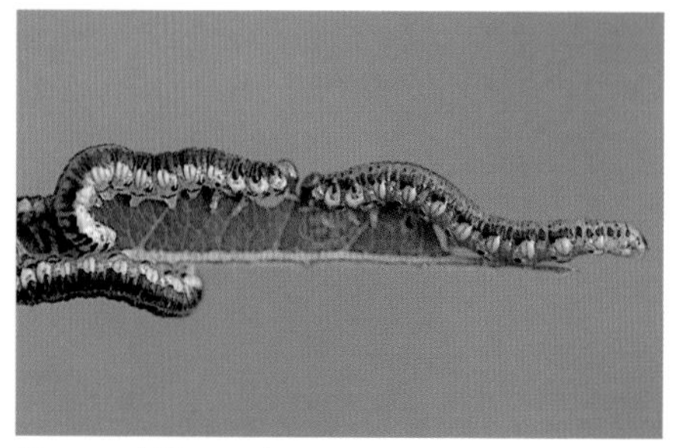

Nematus iridescens, larvae

Pontania californica, galls

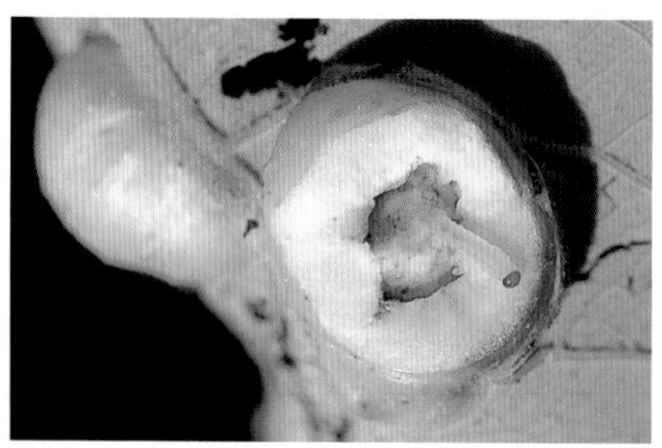

Pontania californica, interior of gall with larva

The female uses her ovipositor to cut slits into leaves into which she deposits her eggs. As the larvae grow, they start feeding in groups along the edges of leaves. When a group of larvae is threatened, the larvae hold onto the leaves with their true legs, wave their rear ends in the air, and, from their anal glands, exude drops of brown liquid used to repel predators and parasites. The larvae may be found feeding on the same plant as the *Nematus iridescens* larvae.

Nematus iridescens

ADULT Shiny green. **BODY LENGTH** 7 mm. **LARVA** Green or pink, variably patterned with black, and with large, yellow, liplike, lateral spiracles. **FOOD** Larva: Coastal willow (*Salix hookeriana*) and other willows (*Salix* spp.). **FOUND** Throughout the region.

The female uses her ovipositor to cut slits into leaves into which she deposits her eggs. As the larvae grow, they start feeding in groups along the edges of leaves. When a group of larvae is threatened, the larvae hold onto the leaves with their true legs, wave their rear ends in the air, and, from their anal glands, exude drops of brown liquid used to repel predators and parasites. The larvae may be found feeding on the same plant as the *Nematus chalceus* larvae.

Pontania californica

WILLOW APPLE LEAF GALL

ADULT Shiny black with brownish red markings. **BODY LENGTH** 4 mm. **LARVA** Small; pale yellow to white; head brown. **GALL** Round; protrudes from both sides of leaf—shiny red above, greenish yellow with wartlike scales below; contains one sawfly larva. **GALL WIDTH** 12 mm. **FOOD** Larva: Coastal willow (*Salix hookeriana*) and other willows (*Salix* spp.). **FOUND** Throughout the region.

TRUE WASPS Family Vespidae

This is a large, common family of social and solitary wasps that are well known for their painful sting. The most infamous group within this family are the social vespids, which includes the paper wasps, yellow jackets, and hornets. Many true wasps are brown or black, but others are banded with yellow, white, or red. True wasps can also be recognized by the fact that their wings are usually folded longitudinally over the abdomen when at rest. Social structure within this family varies from the solitary wasps that construct cells of mud or dig mud-lined underground tunnels for their larvae to those that live in colonies, lay eggs in the combs of cells made of paper, and show some degree of social behavior.

Vespula maculata

WHITE-FACED HORNET

ADULT Black; thorax and abdomen with white patches; face white; eyes black. **BODY LENGTH** 20 mm. **FOOD** Adult: Invertebrates. **FOUND** Throughout the U.S.

Also known as *Dolichovespula maculata*. This is the only white-and-black yellow jacket; all other yellow jackets (*Vespula* spp.) on the West Coast are yellow and black. Overwintering mated females start new colonies in the spring. The nests are above ground in the open. The workers chew wood into pulp that forms the nests and feed the larvae prechewed insects. From early spring through summer, the adults that emerge from the larval stage are workers; in late summer, reproductive males and females emerge and mate. All members of the colonies eventually die, except for the newly mated females, who will start new colonies. This species normally defends its nest very aggressively, and individuals are able to sting multiple times, making them more dangerous than honey bees.

Vespula maculata

BUTTERFLIES AND MOTHS Order Lepidoptera

Lepidoptera (Greek, *lepido*, scale; *ptera*, wings) is the second-largest order of insects, with approximately 112,000 described species worldwide. Over 11,200 known species occur in the United States and Canada. These easily recognizable insects can be found almost everywhere, often in substantial numbers. Although moths are not as familiar to the layperson as butterflies, they have a greater number of species within their ranks and are much more diverse in structural variations and habits than butterflies.

Butterflies and moths range in size from small to large with most in the Pacific Northwest medium-sized. A distinctive characteristic of this order, from which the name is derived, is the two pairs of large, membranous wings, as well as the body and legs, of the adults covered with overlapping scales. The scales, which are modified hairs, give the wings their color and texture. The wings of many species are brightly colored. The forewings are usually larger than the hindwings but structurally similar. At rest, butterflies typically hold their wings together vertically over the body, whereas moths hold their wings either rooflike over the body, curled around the body, or flat against the surface on which they rest. Another distinguishable feature of this order is that the adults of most species have sucking mouthparts that have been modified to form a long, coiled tube, or *proboscis*, which is tucked under the head when not in use. The antennae of butterflies are slender and have an enlarged club at the tip, whereas those of moths are generally serrate or feathery and usually taper to a point.

Lepidopterans undergo complete metamorphosis. Although they greatly vary in shape, the larvae, often called *caterpillars*, usually have a cylindrical, elongate body and a well-developed head with robust jaws. They have three pairs of true legs and up to five pairs of abdominal prolegs. Pupation occurs usually either as a chrysalis (butterflies) or within a silken cocoon or earthen cell (most moths).

Most adult lepidopterans can be found flying around vegetation. Butterflies fly only during the day. Most species of moths fly during the night and are subdued in color, but some are day-flyers and brightly colored. Many adults feed on liquids such as nectar, tree sap, honeydew, animal waste, or decaying organic material, while others do not feed at all. The caterpillars of butterflies and most moths feed on plants. A few moth species eat fungi, lichen, or dried animal matter such as wool or fur, and a few others are predaceous.

The larvae of some lepidopteran species, such as cutworms, codling moths, armyworms, and budworms, are pests of crops, ornamental plants, and stored products; however, the adults of many species are important pollinators of flowering plants. And, because of their conspicuous beauty, butterflies have fascinated humans and played a significant part in mythology, art, and literature, as well as scientific pursuit, throughout the millennia.

Birds are significant predators of butterfly and moth larvae, and, as a result, different strategies have evolved to keep the larvae from being eaten. Three of these methods involve body hair, mimicry, and toxicity.

The larvae of many species of Lepidoptera

have bodies that are covered with hairs. Most birds avoid these hairy larvae because if the larvae are swallowed, their hairs can be irritating to a bird's throat or block a bird's digestive tract. Once a bird has sampled a larva and discovered that it is not palatable, certain characteristics of the larva (for example, coloration) alert the bird to stay away the next time it encounters another larva that is similar to the one it had sampled.

The larvae of some species have soft, relatively short hairs that irritate a bird's throat or stomach if the larvae are swallowed. The tent caterpillar (*Malacosoma* spp.) is an example of a hirsute moth species in this category. The larvae of all three species of tent caterpillars listed in this field guide have characteristics that make them easily identifiable to birds: they are gregarious and brightly colored (the color blue, which all three of these species possess, is not a common color in insects), and two of the three species build tents.

The bodies of the larvae of other species are covered with long, thick, stiff hairs that make it very difficult for a bird to swallow these larvae. The banded woollybear (*Pyrrharctia isabella*) is an example of a larva with this type of physical defense mechanism. Its distinct coloration and pattern is a warning to birds that it is inedible.

Another type of defense is one that is both mechanical and chemical in nature. The larvae of some species, such as the sheep moth (*Hemileuca eglanterina*), have bodies that are covered with urticating spines or hairs that protect them from vertebrate predators. In the late instar stages, the spines of the sheep moth larva give it a very formidable appearance.

Not all birds avoid eating hairy lepidoptera larvae. One predator of the tent caterpillar is the yellow-billed cuckoo (*Coccyzus americanus*), which actively seeks out the larvae, even to the extent of breaking into the tent caterpillars' nests to feed on the larvae inside. The cuckoo is able to ingest the hairy larvae because it has the unique ability to periodically shed its stomach lining and grow a new one. This ability allows the yellow-billed cuckoo to take advantage of a rich food source with little competition from other birds.

Another predator of the tent caterpillar is the Steller's jay (*Cyanocitta stelleri*). Only after the larvae have pupated does this jay eat them. When a larva pupates, it spins a silk cocoon around itself, after which it sheds its skin and hair, leaving only the silken pupal case to protect it. Using their feet and large bills, the jays remove the pupae from the cocoons and eat them. During years when the numbers of tent caterpillars are high, the ground may be littered with empty cocoons.

Another way moth and butterfly larvae avoid being eaten is the use of mimicry. Good examples of this type of defense are the young larvae of most swallowtail butterflies (*Papilio* spp.) and the larvae and pupae of the Lorquin's admiral butterfly (*Basilarchia lorquini*), which look like bird droppings. What object would hold less interest to a hungry bird than its own feces? Bird-dropping mimics are usually hairless, darkly colored with a splash of white across their backs, and often shiny or moist-looking, like a fresh bird dropping.

Finally, some insects protect themselves from bird predation by being poisonous. The monarch (*Danaus plexippus*) and the pipevine swallowtail (*Battus philenor*) are examples of insects that are poisonous as larvae and as adults. Their bodies concentrate poisons they obtain from feeding on their host plants. The larvae and adults of these poisonous insects may also have leathery bodies that help them survive being "tasted" by a bird. The poisons, however, are no protection against parasites and invertebrate predators. In fact, there is evidence that the adult females of some parasitic species actively seek out poisonous larvae, such as that of the monarch, because the larva, as host, provides a safe internal haven from predators for the parasite's eggs and larvae.

Key to Lepidoptera Families

TIGER MOTHS
Family Arctiidae, page 142

OAK MOTHS
Family Dioptidae, page 153

HOOK-TIP MOTHS
Family Drepanidae, page 154

GEOMETER MOTHS
Family Geometridae, page 155

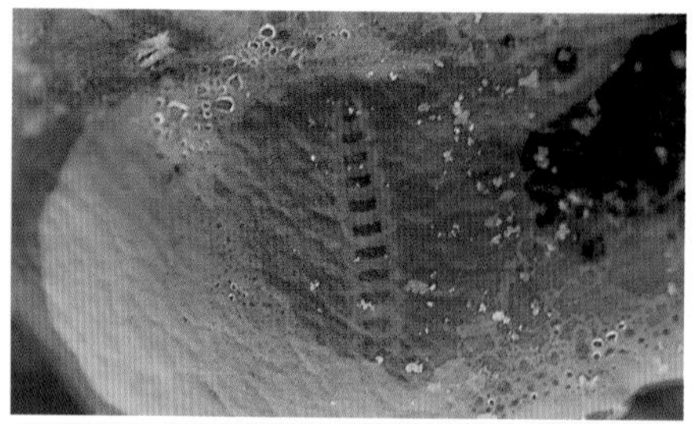

LEAF BLOTCH MINER MOTHS
Family Gracillariidae, page 160

SKIPPERS
Family Hesperiidae, page 161

YUCCA & FAIRY MOTHS
Family Incurvariidae, page 164

LAPPET MOTHS
Family Lasiocampidae, page 165

SLUG CATERPILLAR MOTHS
Family Limacodidae, page 167

HAIRSTREAK, COPPER & BLUE BUTTERFLIES
Family Lycaenidae, page 168

TUSSOCK MOTHS
Family Lymantriidae, page 178

NOCTUID MOTHS
Family Noctuidae, page 180

PROMINENT MOTHS
Family Notodontidae, page 190

BRUSH-FOOTED BUTTERFLIES
Family Nymphalidae, page 193

SWALLOWTAIL & PARNASSIAN BUTTERFLIES
Family Papilionidae, page 213

SULPHUR, WHITE & ORANGE-TIP BUTTERFLIES Family Pieridae, page 221

BAGWORM MOTHS
Family Psychidae, page 225

SNOUT MOTHS
Family Pyralidae, page 226

METALMARK BUTTERFLIES
Family Riodinidae, page 227

GIANT SILKWORM MOTHS
Family Saturniidae, page 228

CLEAR-WINGED MOTHS
Family Sesiidae, page 233

HAWK MOTHS
Family Sphingidae, page 235

THYATIRID MOTHS
Family Thyatiridae, page 240

TORTRICID MOTHS
Family Tortricidae, page 241

TIGER MOTHS Family Arctiidae

These small- to medium-sized moths are often brightly colored with white, yellow, red, or black, and many are boldly patterned with spots or bands. Many species are toxic, and their conspicuous patterning serves as a warning to predators. Tiger moths have stout, furry bodies and broad wings that are held rooflike over the body when at rest. The adults of most species are nocturnal, but some are conspicuous day-flyers. In general, the adults do not feed. The larvae are usually very hairy; the caterpillars known as *woolly-bears* are in this family. Like the adults, they are often boldly marked, and some may be toxic. They feed on a wide range of plants and pupate in cocoons loosely made of body hairs silked together.

Arctia caja

GARDEN TIGER MOTH

ADULT Above, forewings brown with irregular patches of white; hindwings orange with large, blue spots ringed with black. **WINGSPAN** 65 mm. **LARVA** Velvety black; front quarter of body with long, reddish orange hairs; posterior three-quarters of body with long, black and pale yellow hairs on back and long, reddish orange hairs on sides. **FOOD** Larva: Mainly herbaceous plants such as English plantain (*Plantago lanceolata*) and other plantains (*Plantago* spp.), lupine (*Lupinus* spp.), and the dandelion *Taraxacum officinale*. **FOUND** Widely scattered throughout the region. Not always easy to find but well worth the effort. Once located, can generally be found at the same site year after year. Look for mature larvae crossing roads in early spring.

The adult flies in midsummer. When physically threatened, the adult produces a liquid from glands at the base of its head.

Arctia caja

Arctia caja, larva

Cisseps fulvicollis

YELLOW-COLLARED SCAPE MOTH

ADULT Above, forewings brown; body black with orange area posterior to head. Similar in shape (wasplike) to *Ctenucha multifaria*. **WINGSPAN** 35 mm. **FOOD** Larva: Grasses and grasslike plants. **FOUND** Throughout the region. Can find adults on flowers in late summer.

Ctenucha multifaria

ADULT Wings black, often with white margins; abdomen iridescent blue; head and shoulders red. Similar in shape (wasplike) to *Cisseps fulvicollis*. **WINGSPAN** 45 mm. **LARVA** Black covered with pale brown hairs, and black tufts at anterior and posterior ends. **FOOD** Larva: Grasses and grasslike plants. **FOUND** Throughout the region.

Pete used the key in Essig (1958) to identify this species; some entomologists refer to this species as *Ctenucha rubroscapus*, and others think that *C. multifaria* and *C. rubroscapus* are one and the same. This species flies during the day.

Cisseps fulvicollis

Ctenucha multifaria, mating pair, male on right

Ctenucha multifaria, larva

Estigmene acrea

Estigmene acrea, larva

Gnophaela latipennis

Gnophaela latipennis, larva

Estigmene acrea

SALTMARSH CATERPILLAR

ADULT Above, forewings white with black spots; hindwings in male brownish yellow and in female white. **WINGSPAN** 64 mm. **LARVA** Black with pale yellow stripes on sides; back covered with black hairs, sides with red hairs. **FOOD** Larva: Herbaceous plants such as plantain (*Plantago* spp.), lupine (*Lupinus* spp.), clover (*Trifolium* spp.), and grasses. **FOUND** Throughout most of the U.S.

Adult males have abdominal glands that release chemicals that attract both females and males, resulting in mating aggregations of adults of both sexes.

Gnophaela latipennis

SIERRAN PERICOPID

ADULT Above, wings black with creamy white patches. **WINGSPAN** 54 mm. **LARVA** Yellow with black lines and spots; covered with tufts of white hairs; head red. **FOOD** Larva: The hound's tongue *Cynoglossum grande* and related plants. **FOUND** Throughout the region.

This day-flying moth is a weak flyer; it can often be caught by hand when it is nectaring on flowers.

Grammia ornata

ORNATE TIGER MOTH

ADULT Above, forewings black with creamy pink crosshatching; hindwings red-orange with black spotting; abdomen light orange with dorsal row of black spots. **WINGSPAN** 40 mm. **LARVA** Black, covered with black hairs and with fringe of light red hairs. **FOOD** Larva: Mainly herbaceous plants such as English plantain (*Plantago lanceolata*) and other plantains (*Plantago* spp.), lupine (*Lupinus* spp.), and the dandelion *Taraxacum officinale*. **FOUND** Throughout the region. May be very common, but larva is very difficult to find in

Grammia ornata

vegetation. Look for early instar larvae crossing roads in fall and mature larvae crossing roads in early spring. Adult male sometimes attracted to lights at night.

When threatened, the adult releases a liquid from the back of its head that acts as a repellent to predators.

Hemihyalea edwardsii

EDWARDS' GLASSYWING

ADULT Above, forewings translucent, pale orangish brown with faint black markings; abdomen pinkish orange with black spots. **WINGSPAN** 60 mm. **LARVA** Covered with stiff, shiny, black and reddish brown hairs. **FOOD** Larva: Oak (*Quercus* spp.). **FOUND** Oregon and California.

The wings of the adult lose their scales soon after the adult's first flight, giving the wings their translucent quality.

Grammia ornata, larva

Hemihyalea edwardsii

Hemihyalea edwardsii, larva

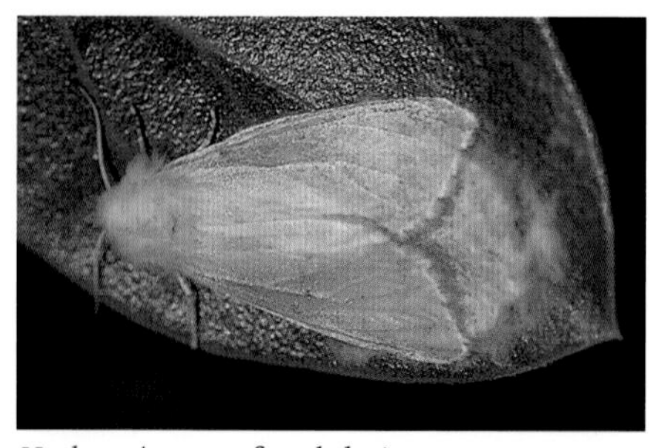

Hyphantria cunea, female laying eggs

Hyphantria cunea, larva

Hyphantria cunea, nest

Hyphantria cunea

FALL WEBWORM

ADULT Above, forewings usually completely white, occasionally with black spots; thorax covered with white hairs; front femurs with orange hairs. **WINGSPAN** 34 mm. **LARVA** Covered with long white to pale brown hairs arising from orange bumps. **FOOD** Larva: Trees and shrubs such as Pacific madrone (*Arbutus menziesii*), red alder (*Alnus rubra*), and Oregon ash (*Fraxinus latifolia*). **FOUND** Throughout the U.S.

This species tends to be cyclic, building up to high numbers then seemingly disappearing for a few years. Although its nest looks similar to that of a tent caterpillar, it occurs at a different time of the year: it is most common from late summer to fall, whereas the tent caterpillar larva is usually most common from spring to midsummer. The larvae build a silken nest within which they reside and feed. The nest gets progressively larger and fills up with their feces and defoliated branches as the larvae grow and their consumption of food increases.

Leptarctia californiae

Leptarctia californiae

Leptarctia californiae, larva

Leptarctia californiae

CALIFORNIA TIGER MOTH

ADULT Above, forewings black, each with one or two white spots on leading margin; hindwing color and pattern extremely variable—usually black with orange or red areas; head black with pale pink lateral stripes. **WINGSPAN** 30 mm. **LARVA** Mottled gray with pale white dorsal stripe; covered with black hairs, those on posterior end long. **FOOD** Larva: Woody plants (e.g., California-lilac [*Ceanothus* spp.]) and herbaceous plants (e.g., peregrine thistle [*Cirsium cymosum*]). **FOUND** Common throughout the region, but larva secretive and adult well camouflaged.

The larva has a very long life cycle, often lasting from spring to late summer.

Lophocampa argentata

Lophocampa argentata, larva

Lophocampa argentata

SILVER-SPOTTED TIGER MOTH

ADULT Above, forewings dark brown with white spots. **WINGSPAN** 40 mm. **LARVA** Black with dorsal tufts of black and yellow hairs; brown or reddish brown hairs laterally. **FOOD** Larva: Native and introduced conifers including Douglas-fir (*Pseudotsuga menziesii*), Sitka spruce (*Picea sitchensis*), and lodgepole pine (*Pinus contorta*). **FOUND** Common throughout the region. Larva usually feeds in groups and is easy to spot: look for branches stripped of their needles.

Also known as *Halisidota argentata*.

Lophocampa maculata

YELLOW-SPOTTED TIGER MOTH

ADULT Above, forewings dark golden brown with light yellow spots that often coalesce, giving the wing a banded appearance. **WING-**

Lophocampa maculata

Lophocampa maculata, larva, typical pattern

Lophocampa maculata, larva

Lophocampa maculata, larva

SPAN 45 mm. **LARVA** Extremely variable in coloration and pattern (three variations shown); most common variation—anterior and posterior ends of body covered with black hairs; middle of body covered with reddish brown hairs; tufts of long, white hairs over entire body. Similar to *Pyrrharctia isabella* except that latter lacks tufts of long, white hairs. **FOOD** Larva: Coastal willow (*Salix hookeriana*) and other willows (*Salix* spp.), red alder (*Alnus rubra*), and other broadleaf trees. **FOUND** Common throughout the region. Larva found in the fall on the ground, commonly crossing roads, looking for places to pupate and overwinter.

Also known as *Halisidota maculata*.

Platyprepia virginalis

RANGELAND TIGER MOTH

ADULT Above, forewings black and with many large, white to creamy white spots; hindwings variably patterned (three variations shown) but usually either black with white or orange spots or orange with black spots. **WINGSPAN** 60 mm. **LARVA** Large; variably patterned with black, orangish red, and light brown hairs; with dorsal mane of longer white hairs. **FOOD** Larva: Mainly herbaceous plants such as English plantain (*Plantago lanceolata*) and other plantains (*Plantago* spp.), lupine (*Lupinus* spp.), and the dandelion *Taraxacum officinale*. **FOUND** Throughout the region, from the coast to the mountains. Adult commonly flies during the day in midsummer. Larva commonly occurs on roadside weeds and is often

Platyprepia virginalis

Platyprepia virginalis

Pyrrharctia isabella

Pyrrharctia isabella, larva

Platyprepia virginalis, larva

seen crossing roads in the spring looking for a safe place to pupate.

Also known as *Platyprepia guttata*. This is one of the most common and largest woolly-bears in the region.

Pyrrharctia isabella

BANDED WOOLLYBEAR

ADULT Above, forewings golden brown with few to many small, black spots. Front femurs orange-brown. **WINGSPAN** 50 mm. **LARVA** Densely covered with stiff shiny hairs—black hairs on anterior and posterior ends and broad band of brownish red hairs around middle. **FOOD** Larva: Mainly herbaceous plants such as English plantain (*Plantago lanceolata*) and other plantains (*Plantago* spp.), lupine (*Lupinus* spp.), and the dandelion *Taraxacum officinale*. **FOUND** Throughout much of the U.S. Larva is commonly seen in late fall and early spring crossing roads. In the fall, it is searching for a protected spot in which to overwinter, and, in the spring, a place to pupate. The adult emerges early to midsummer.

Also known as *Isia isabella*. This species is the woollybear caterpillar of the old wives' tale: according to this tale, one can predict the severity of the coming winter by the width of the brownish red band around the middle of the caterpillar's body—the wider the band, the colder the winter.

Spilosoma vagans

WANDERING TIGER MOTH

ADULT Above, male, forewings gray-brown, sometimes with small, black spots or other markings; female, forewings very rich reddish brown, sometimes with small, black spots or other markings. **WINGSPAN** 38 mm. **LARVA** With broad dorsal band of black hairs; reddish brown hairs laterally. **FOOD** Larva: Plants. **FOUND** Throughout the region. Pete has found this species in such diverse habitats as mountain meadows and weedy city lots.

Spilosoma vagans, female and male

Spilosoma vestalis

VESTAL TIGER MOTH

ADULT Above, forewings white with few to many small black spots; abdomen with rows of black spots; front femurs red. **WINGSPAN** 50 mm. **LARVA** With broad dorsal band of black hairs; reddish brown hairs laterally. **FOOD** Larva: Mainly herbaceous plants such as English plantain (*Plantago lanceolata*) and other plantains (*Plantago* spp.), lupine (*Lupinus* spp.), and the dandelion *Taraxacum officinale*. **FOUND** Throughout the region.

This is probably the most common *Spilosoma* species in the region.

Spilosoma vagans, larva

Spilosoma vestalis, larva

Spilosoma vestalis

Spilosoma virginica

Spilosoma virginica, larva

Spilosoma virginica

YELLOW WOOLLYBEAR

ADULT Above, forewings usually pure white or white with few, small, black spots; abdomen white with black spots and with some orange on dorsal side; front femurs gold or pale yellow. **WINGSPAN** 50 mm. **LARVA** Covered with bright yellow (common) or dull brownish yellow (uncommon) hairs. **FOOD** Larva: Many herbaceous and woody perennial plants such as willow (*Salix* spp.) and *Rubus* spp. **FOUND** Throughout the U.S.

Also known as *Diacrisia virginica*.

OAK MOTHS Family Dioptidae

Most species in this small family of moths are brightly marked, resembling some butterflies; however, the adults of the only two species that occur in the United States (only one in the Pacific Northwest) have wings that are pale brown with dark veins.

Phryganidia californica

CALIFORNIA OAK MOTH

ADULT Above, forewings pale brown, each with light yellow patch in middle. **WINGSPAN** 40 mm. **LARVA** Complex pattern of broken black lines with yellow, black, and white areas. **EGG** White with red cap; laid in groups. **FOOD** Larva: Oak (*Quercus* spp.), tan(bark) oak (*Lithocarpus densiflorus*). **FOUND** Oregon and California; very common in western California.

Pete has seen large concentrations of adults but no serious larval damage in areas of native vegetation. The larvae can devastate ornamental plantings.

Phryganidia californica

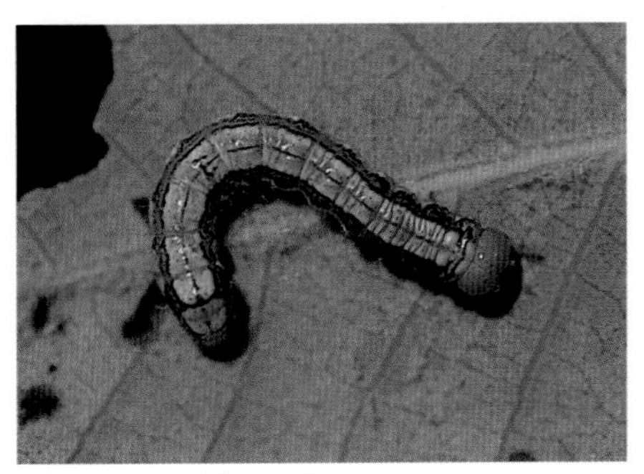

Phryganidia californica, larva

HOOK-TIP MOTHS Family Drepanidae

The members of this small family of moths can generally be recognized by the sickle-shaped tip of the forewings, hence the common name. They are small to medium in size, slender-bodied, broad-winged, and usually dull-colored. The larvae feed on the foliage of a variety of plants.

Drepana arcuata

Drepana arcuata, larva

Drepana arcuata

ARCHED HOOK-TIP MOTH

ADULT Above, forewings light brown with thin, scalloped, dark brown lines and each with one medium brown band ending at hooked tip. **WINGSPAN** 35 mm. **LARVA** Green with brown on back and two rows of four bumps posterior to head; body tapers posteriorly. **FOOD** Larva: Red alder (*Alnus rubra*) and other alders (*Alnus* spp.) **FOUND** Throughout the region. Look for larva inside partially closed leaf that is silked together.

GEOMETER MOTHS Family Geometridae

This is the third-largest family of lepidopterans. They can be easily recognized by their habit, when at rest, of spreading their wings out flat against the substrate with both the forewings and hindwings showing. The adults of most species are small to medium in size, delicate, and slender-bodied; the wings are usually broad and often marked with fine wavy lines. In a few species, the females are wingless. Most are nocturnal and often attracted to lights. Some adults feed, others do not. The larvae, commonly called *inchworms* or *measuringworms*, are slender caterpillars that have one or two pairs of abdominal prolegs and are easily recognizable by the way they move—they draw the posterior part of the body up to the thorax and then extend the anterior part of the body as far forward as possible, thus progressing in a characteristic looping manner. At rest or when disturbed, the larvae of some species stand almost erect on their posterior prolegs and remain motionless, resembling twigs. Others attach pieces of plant material to their back to conceal themselves. The larvae feed on many types of plants, usually on the leaves. Most pupate in loose cocoons in leaf litter or soil. Although many species are common, it is difficult to identify this family to species.

Biston betularia

PEPPER-AND-SALT MOTH

ADULT Wings pale gray with wavy, black lines and heavily speckled with dark gray spots. **WINGSPAN** 48 mm. **LARVA** Twig mimic; brown to green; head bilobed. **LARVA LENGTH** 60–75 mm. **FOOD** Larva: Mostly deciduous trees and shrubs such as apple (*Malus* spp.), willow (*Salix* spp.), and maple (*Acer* spp.). **FOUND** Common throughout the U.S.

This moth has been used as a case study to illustrate the phenomenon called *industrial melanism*. Industrial air pollution in England was linked to the change in the color of this species from the typical, light salt-and-pepper to a much darker form. In areas where industrial air particulates settled over the landscape, the light-colored adults were much more visible than the dark-colored adults and hence were more likely to be eaten by birds. When moths were collected from these areas prior to 1848, researchers found that the light-colored form made up 99 per-

Biston betularia

Biston betularia, larva

cent of the population; in a complete reversal, collections done in 1898 and later showed that 99 percent of the population was composed of the dark-colored form. See Kettlewell (1973) for details of this fascinating story and Hooper (2002) for a critical look at Kettlewell's work.

Campaea perlata

ADULT Above, forewings pale green, each with two white lines; hindwings pale green, each with one white line. **WINGSPAN** 39 mm. **LARVA** Mottled brown with lateral fringe of "hairs." **FOOD** Larva: Woody broadleaf trees and shrubs including willow (*Salix* spp.). **FOUND** Throughout the region. Common but both larva and adult well camouflaged: the larva's mottled appearance makes it difficult to see on bark, and the adult's wings blend into green foliage. To find larvae, first look for larval feeding activity on leaves, which normally occurs at night, then for larvae on large branches or at base of host plant during the day.

Campaea perlata

Campaea perlata, larva

Cochisea sinuaria

Cochisea sinuaria

ADULT Above, wings gray with black lines and lightly speckled with black spots. **WINGSPAN** 60 mm. **LARVA** Twig mimic; reddish brown (same color as the young stems of madrone). **FOOD** Larva: Pacific madrone (*Arbutus menziesii*), manzanita (*Arctostaphylos* spp.). **FOUND** Oregon and California.

The adult emerges in late summer to fall.

Cochisea sinuaria, larva

Erannis tiliaria

LINDEN LOOPER

ADULT Male, forewings brown, each with large pale white area in center; female, black, wingless. **WINGSPAN** 40 mm. **LARVA** Dark brown; broad, yellow lateral band with reddish brown patches. **FOOD** Larva: Bigleaf maple (*Acer macrophyllum*) and other maples (*Acer* spp.), California black oak (*Quercus kelloggii*) and other oaks (*Quercus* spp.). **FOUND** Throughout the U.S.

The adult emerges in winter. The larva assumes a unique position when disturbed—its body is hunched and its head and first two pairs of true legs are elevated (see photograph).

Erannis tiliaria, female

Erannis tiliaria, larva

Mesoleuca gratulata

ADULT Above, forewings mostly black, each with wide, white band containing black spot in middle. **WINGSPAN** 22 mm. **FOOD** Larva: The hazelnut *Corylus cornuta* (leaves). **FOUND** Throughout the region. Common in early spring.

Mesoleuca gratulata

Nemoria darwiniata

Nemoria darwiniata, larva

Neoterpes trianguliferata

Pero occidentalis

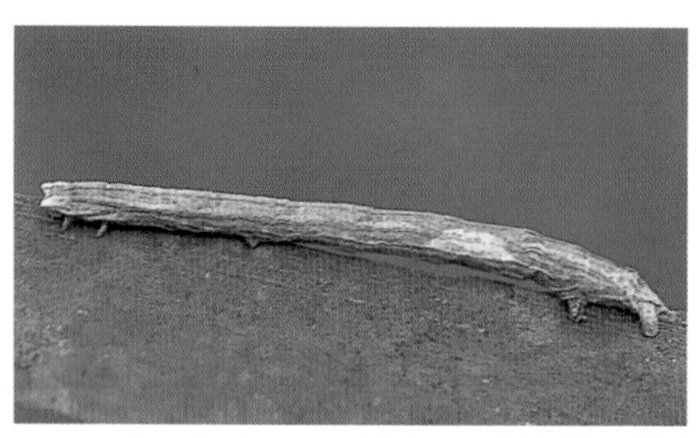
Pero occidentalis, larva

Nemoria darwiniata

ADULT Above, wings green, each with two white medial lines and tiny, red spot in middle; abdomen green with white spots ringed with red. **WINGSPAN** 30 mm. **LARVA** Light and dark brown with three dorsal flaps. **FOOD** Larva: Willow (*Salix* spp.), California-lilac (*Ceanothus* spp.), and many other trees and shrubs. **FOUND** Common throughout the region. Adult attracted to lights at night. Larva is so well camouflaged that it is extremely hard to find on host plant.

Neoterpes trianguliferata

ADULT Above, forewings yellow, each with two brown bars extending inward from outer margin. **WINGSPAN** 30 mm. **FOOD** Larva: Currant or gooseberry (*Ribes* spp.) (leaves). **FOUND** Common throughout the region. Adult attracted to lights at night.

Pero occidentalis

ADULT Above, forewings gray-reddish brown with scalloped outer margins and dark reddish brown medial band. **WINGSPAN** 34 mm. **LARVA** Pattern and color extremely variable but usually gray with light black reticulations. **FOOD** Larva: Conifers. **FOUND** Throughout the region.

Sabulodes aegrotata

Sabulodes aegrotata, larva

Sabulodes aegrotata

OMNIVOROUS LOOPER

ADULT Above, wings pale brown with fine, dark brown to black speckling and dark brown to black transverse bands. **WINGSPAN** 40 mm. **LARVA** Yellow and green with black and white longitudinal lines, additionally often with orange or brown lines. **FOOD** Larva: Plants. **FOUND** Throughout the region. Larva commonly feeds on ornamental plantings.

The larva silks together leaves, within which it spends the day; it emerges at night to feed. The adult flies year round in mild winter areas.

Sicya crocearia

Sicya crocearia

ADULT Above, forewings yellow, each with large, pinkish brown patch on outer margin and two pinkish brown medial lines. **WINGSPAN** 35 mm. **LARVA** Green with two large, reddish brown dorsal protuberances and one smaller, reddish brown dorsal protuberance near posterior end. **FOOD** Larva: Red alder (*Alnus rubra*) and other alders (*Alnus* spp.). **FOUND** Throughout the region.

Sicya crocearia, larva

Spargania magnoliata

Spargania magnoliata

ADULT Above, forewings various shades of gray with gold flecks and scalloped, dark gray and black lines. **WINGSPAN** 20 mm. **FOOD** Larva: *Epilobium brachycarpum* and other fireweeds (*Epilobium* spp.). **FOUND** Common throughout the region.

LEAF BLOTCH MINER MOTHS Family Gracillariidae

The members of this large family are minute to small moths with wings that are long and narrow (the hindwings are very narrow), often with long fringes. The adults at rest often elevate the front of their body and lay the tips of their folded wings on the surface on which they rest. The adults are very difficult to identify; knowing the species of the host plant on which one finds the moth is very helpful as an aid in identifying the moth species. The family gets its common name from the larvae, which usually make blotch mines in leaves.

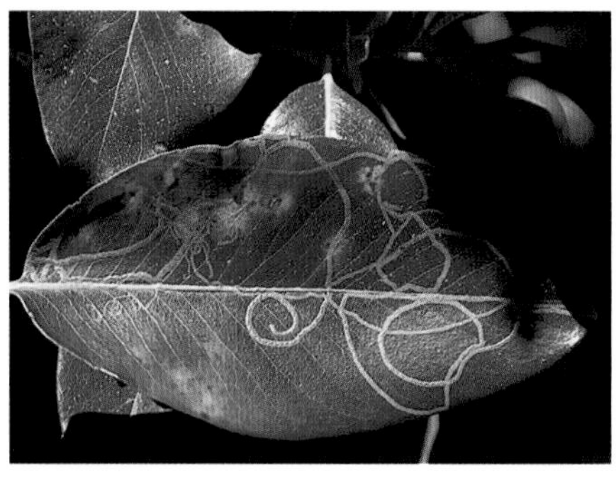

Marmara arbutiella, larva mining leaf

Phyllonorycter nemoris, larva within leaf

Marmara arbutiella

MADRONE SKIN MINER

ADULT Black; forewings with thin, white lines. **WINGSPAN** 3 mm. **LARVA** Tiny; pale brown. **FOOD** Larva: Pacific madrone (*Arbutus menziesii*) (leaves). **FOUND** Range unknown but probably same as host.

The larva's feeding activity creates a pattern ("mining") on the leaf. The site on the leaf where the mining is very narrow is where the larva first starts feeding. As the larva grows, the mining gradually gets larger (the photograph shows the larva at the end of the mining in the middle of the leaf).

Phyllonorycter nemoris

HUCKLEBERRY LEAF BLOTCH MINER

ADULT Tiny; above, forewings orange with white bands. **WINGSPAN** 8 mm. **LARVA** Pale reddish brown with black dorsal bands. **FOOD** Larva: California huckleberry (*Vaccinium ovatum*) (leaf miner). **FOUND** Throughout the region. Larva feeds under epidermis of leaf: look for leaf with separated upper and lower layers and the larva and its dark excrement between the two surfaces.

Also known as *Cameraria nemoris*.

SKIPPERS Family Hesperiidae

The members of this large family get their common name from their characteristic fast, darting flight. They differ from other butterflies in that they have proportionately larger bodies, smaller wings, and antennae tips that are usually hooked as well as clubbed. For the most part they are small- to medium-sized with stout, often hairy bodies and triangular wings. They are usually a dull brown, often with orange or white markings. Most of the small-winged species hold their forewings at an angle different from the hindwings when at rest. The larvae of most species are usually smooth-skinned and have a large head and constricted neck. They typically feed inside silked-together leaf shelters. Most overwinter as larvae or pupae, often in loose cocoons.

Carterocephalus palaemon

ARCTIC SKIPPER

ADULT Above, wings dark brown with distinct, angular, golden yellow spots. Below, hindwings reddish brown with angular, pale yellow spots. **WINGSPAN** 32 mm. **FOOD** Adult: Flowering plants (nectar). Larva: Grasses. **FOUND** Throughout the northern U.S.; on the West Coast, as far south as central California. Most common at high elevations but may also occur in lowlands.

This is one of the easiest skippers to identify.

Carterocephalus palaemon

Epargyreus clarus

SILVER-SPOTTED SKIPPER

ADULT Above, forewings dark brown, each with orange band through middle; hindwings dark brown. Below, forewings dark brown; hindwings dark brown, each with white band through middle. **WINGSPAN** 60 mm. **LARVA** Greenish yellow with fine, black bands; head dark brown with two orange spots. **FOOD** Adult: Flowering plants (nectar). Larva: Locust (*Robinia* spp.), wisteria (*Wisteria* spp.), honey locust (*Gleditsia* spp.). **FOUND** Throughout the region.

This is one of the largest skippers in the region.

Epargyreus clarus

Epargyreus clarus, larva

Erynnis propertius

Erynnis tristis

Erynnis tristis, larva

Erynnis propertius

PROPERTIUS DUSKYWING

ADULT Above, forewings with patches and spots of various shades of gray and brown; hindwings brown with light brown spots. **WINGSPAN** 44 mm. **FOOD** Adult: Flowering plants (nectar). Larva: Tan(bark) oak (*Lithocarpus densiflorus*), oak (*Quercus* spp.). **FOUND** Throughout the region. Usually the first duskywing species seen in the spring in oak woodlands.

Erynnis tristis

MOURNFUL DUSKYWING

ADULT Above, forewings mottled dark brown, each with row of four small, translucent spots off leading margin; hindwings dark brown with white fringe. Similar to other duskywings but hindwing fringe sets it apart. **WINGSPAN** 40 mm. **LARVA** Green with pale yellow speckling; head brownish red. **FOOD** Adult: Flowering plants (nectar). Larva: Tan(bark) oak (*Lithocarpus densiflorus*), oak (*Quercus* spp.). **FOUND** Southern Oregon and California.

Ochlodes sylvanoides

WOODLAND SKIPPER

ADULT Above, wings orange with dark brown border. Below, wings yellow-brown with pale yellow, angular spots. **WINGSPAN** 28 mm. **FOOD** Adult: Flowering plants (nectar). Larva: Grasses. **FOUND** Throughout the region. Adults appear in large numbers in the fall nectaring at late-blooming flowers such as aster (*Aster* spp.) and goldenrod (*Solidago* spp.).

Ochlodes sylvanoides

Ochlodes sylvanoides

Pyrgus communis

COMMON CHECKERED SKIPPER

ADULT Above, wings dark brown with few to many white spots and blue-gray hairs at base. **WINGSPAN** 32 mm. **LARVA** Light brown; head black; area posterior to head dark brown with two black spots. **FOOD** Adult: Flowering plants (nectar). Larva: Checker mallow (*Sidalcea malvaeflora*) and other checker mallows (*Sidalcea* spp.). **FOUND** Common throughout the U.S.; occurs in gardens. Larva can be found inside one or more leaves silked together.

This species may be the most common skipper in the U.S.

Pyrgus communis, larva

Pyrgus communis

YUCCA & FAIRY MOTHS Family Incurvariidae

The adults in this family are small moths with threadlike antennae that range in length from short to very long; the males of some species have antennae that may be several times the length of the wings, the longest antennae of any Lepidoptera (the members of the genus *Adela* are commonly called *long-horn moths*). Many species are black or metallic blue, sometimes with gold or silver markings. The wings are rounded at the tip and covered with tiny hairlike spines. The females have a piercing ovipositor. This family includes both day-flying and night-flying species. The larvae feed on the foliage or seeds of plants; many are leaf miners, whereas others form protective cases from leaf fragments.

Adela septentrionella

Adela septentrionella

ADULT Forewings black with white spots and lines; antennae long (in males, much longer than body). **WINGSPAN** 12 mm. **FOOD** Larva: Oceanspray (*Holodiscus discolor*). **FOUND** Throughout the region. Adults hover around oceanspray in spring and early summer.

The genus *Adela* is easy to identify; however, the species within this genus are difficult to differentiate from each other. *Adela septentrionella* is the exception to the rule: it can be easily identified because it is the only species of *Adela* that uses a perennial plant for its host.

LAPPET MOTHS Family Lasiocampidae

The adult moths of this mainly tropical family have, in general, medium-sized, stout, hairy bodies. Most species are dull brown or gray in color. The antennae are feathery in both sexes but more so in the males. The wings are held rooflike over the body when at rest. The adults do not feed. The larvae are slender, colorful, and hairy. The common family name refers to the fact that the larvae of some species have small lobes, or lappets, on each side of the body. The larvae feed on the foliage of trees and shrubs. Some species are known as *tent caterpillar moths* because the larvae build communal webs ("tents") on woody plants. These tents serve as protection from predators, although the larvae leave them when feeding. Three tent caterpillar moth species occur in the Pacific Northwest. Pupation in all lappet moth species occurs in well-formed cocoons. Some species are considered to be pests in orchards, where they may completely strip tree branches of their leaves.

Malacosoma californica

WESTERN TENT CATERPILLAR

ADULT Above, forewings reddish brown, each with darker brown band in middle. **WINGSPAN** 28 mm. **LARVA** Dark brown and covered with reddish brown hairs; with white, oblong dorsal patches and two blue spots per segment. **FOOD** Larva: Blue blossom (*Ceanothus thyrsiflorus*) and other California-lilacs (*Ceanothus* spp.); Oregon oak (*Quercus garryana*), California black oak (*Q. kelloggii*), and other oaks (*Quercus* spp.); and many other trees and shrubs. **FOUND** Throughout the region.

Malacosoma californica, larvae on nest

Malacosoma constricta

PACIFIC TENT CATERPILLAR

ADULT Above, forewings light brown, each with two distinct, brown lines through middle. **WINGSPAN** 28 mm. **LARVA** With large, orange dorsal spots surrounded by blue, and lateral tufts of white hairs. **FOOD** Larva: Oregon oak (*Quercus garryana*), California black oak (*Q. kelloggii*), and other oaks (*Quercus* spp.); blue blossom (*Ceanothus thyrsiflorus*) and other California-lilacs (*Ceanothus* spp.); and many other trees and shrubs. **FOUND** Throughout the region.

Malacosoma constricta, larva

Malacosoma disstria

Malacosoma disstria, larva

Phyllodesma occidentis

Phyllodesma occidentis, larva

Malacosoma disstria

FOREST TENT CATERPILLAR

ADULT Above, forewings yellowish to pale brown, each with two brown bands through middle. **WINGSPAN** 29 mm. **LARVA** With dorsal row of white spots surrounded by blue and black; broad, blue lateral band with fine, yellow-orange longitudinal lines. **FOOD** Larva: Alder (*Alnus* spp.); coastal willow (*Salix hookeriana*) and other willows (*Salix* spp.); Oregon oak (*Quercus garryana*), California black oak (*Q. kelloggii*), and other oaks (*Quercus* spp.); blue blossom (*Ceanothus thyrsiflorus*) and other California-lilacs (*Ceanothus* spp.); and many other trees and shrubs. **FOUND** Throughout the U.S.

This is the most common tent caterpillar on the coast, where willow and alder are its favorite foods.

Phyllodesma occidentis

ADULT At rest, looks like dead leaf. Above, forewings reddish brown; hindwings reddish brown with scalloped, white outer margin. **WINGSPAN** 35 mm. **LARVA** White and gray-black with lateral tufts of gray hairs; two reddish orange, transverse, anterior bands displayed when larva is threatened (if further threatened, larva rises up and waves anterior portion of body, exposing brightly colored and patterned underside). **FOOD** Larva: Red alder (*Alnus rubra*), California-lilac (*Ceanothus* spp.), and probably many other woody broadleaf plants. **FOUND** Throughout the region.

SLUG CATERPILLAR MOTHS Family Limacodidae

The adult members of this moth family are small to medium in size, with stout, often hairy bodies and broad, rounded wings. Most are brown with contrasting markings on the wings. The adults do not feed. The family gets its common name from the larvae, which are short, fleshy, and sluglike in appearance and crawl about in a sluglike manner. Many are conspicuously marked. Some species have tufts of short, stinging hairs that protect them from predators; these hairs can be irritating to human skin if the larva is handled. The larvae feed on a variety of woody and herbaceous plants.

Tortricidia testacea

ADULT Above, forewings pale brown. **WINGSPAN** 24 mm. **LARVA** Green with brown dorsal band outlined in red and tapered at both ends. **FOOD** Larva: Bigleaf maple (*Acer macrophyllum*) and other plants. **FOUND** Throughout the region.

Tortricidia testacea

Tortricidia testacea, larva

HAIRSTREAK, COPPER & BLUE BUTTERFLIES

Family Lycaenidae

This family of butterflies is very large and worldwide in distribution. In general, the adults are small, delicate, and, in the males, often with brilliant blue- or copper-colored wings. Many have tiny taillike projections on the margins of the hindwings. These butterflies are rapid flyers, and, when at rest, they hold their wings together over their back, unlike their close relatives, the metalmarks (Riodinidae). The adults nectar on flowers and can also be found puddling (feeding on moist or wet soils). The larvae are flattened, short, oval, and sluglike. Most feed on plants. The larvae of some species are often tended to by ants for the honeydew the larvae produce. The pupae, most of which are oval in shape, are usually attached to a surface by a silk girdle above the ground or in leaf litter.

Atlides halesus

Callophrys dumetorum

Atlides halesus

GREAT PURPLE HAIRSTREAK

ADULT Above, wings iridescent blue with black margins. Below, forewings dark gray with purplish highlights, each with large, red spot at base; hindwings same as forewings but each with two large, red spots at base and blue spots near tail. Abdomen red. **WINGSPAN** 35 mm. **FOOD** Adult: Flowering plants (nectar). Larva: Mistletoe (*Phoradendron* spp.). **FOUND** Oregon and California.

Callophrys dumetorum

BRAMBLE HAIRSTREAK

ADULT Above, wings brown; below, wings bluish green with white fringe. **WINGSPAN** 28 mm. **LARVA** Often pink, less commonly green. **FOOD** Adult: Flowering plants (nectar). Larva: Wild buckwheat (*Eriogonum* spp.), *Lotus* spp. **FOUND** Throughout the region.

Some lepidopterists consider the populations on the coast of California (where the photograph was taken) to be a separate species, *Callophrys viridis*.

Celastrina argiolus

SPRING AZURE

ADULT Below, wings pale gray; hindwings with row of black spots and chevrons along outer margin. **WINGSPAN** 30 mm. **FOOD** Adult: Flowering plants (nectar). Larva: American dogwood (*Cornus sericea*) as well as many other trees and shrubs. **FOUND** Throughout the region.

This species is possibly the most widespread butterfly in the region. The adults congregate in large numbers to puddle.

Celastrina argiolus

Euphilotes enoptes

DOTTED BLUE

ADULT Above, male, wings bright blue with black margins; female, wings dark brown. Below, forewings gray with black spots; hindwings gray with black spots, one row of orange spots submarginally; fringe on wings checkered with white and black. **WINGSPAN** 25 mm. **FOOD** Adult: Flowering plants (nectar). Larva: *Eriogonum nudum* and other wild buckwheats (*Eriogonum* spp.). **FOUND** Throughout the region. Adult usually stays very close to larval host plant.

See Glassberg (2001) and Pyle (1981) for a discussion of this species and *Euphilotes battoides,* which is not described in this guide.

Euphilotes enoptes

Glaucopsyche lygdamus

SILVERY BLUE

ADULT Above, male, wings iridescent blue with narrow, black margins; female, wings brown with some blue near base. Below, wings light brownish gray, each with one row of black spots ringed with white. **WINGSPAN** 32 mm. **FOOD** Adult: Flowering plants (nectar). Larva: *Lupinus rivularis* and other lupines (*Lupinus* spp.) (flowers). **FOUND** Throughout the region, from sea level to high mountains.

The adult is one of the first blues to appear in the spring.

Glaucopsyche lygdamus

Habrodais grunus

Icaricia acmon, male

Icaricia acmon

Icaricia acmon, larva attended by ants

Habrodais grunus

GOLDEN HAIRSTREAK

ADULT Above, wings brown. Below, forewings golden brown, each with postmedial line; hindwings golden brown, each with postmedial line and often with less distinct, silver, submarginal crescents and spots. Hindwings with very short tail. **WINGSPAN** 32 mm. **LARVA** Green with pale yellow lines and covered with pale white hairs. **FOOD** Adult: Flowering plants (nectar). Larva: Canyon live oak (*Quercus chrysolepis*) and other oaks (*Quercus* spp.), tan(bark) oak (*Lithocarpus densiflorus*). **FOUND** Washington (rare), Oregon, and California. Adult is usually on leaves of host plant. Pete has seen very large populations of adults in tan(bark) oak woodlands. Ragged-looking young foliage on host plants is often an indication of larval feeding.

The adult is easy to approach.

Icaricia acmon

ACMON BLUE

ADULT Above, male, forewings iridescent blue, hindwings iridescent blue with orange band; female, forewings brown suffused with blue, hindwings brown suffused with blue and with orange band. Below, forewings light gray with black spots; hindwings light gray with black spots and outer margin with row of black-capped orange crescents. **WINGSPAN** 25 mm. **LARVA** Green with pale yellow, scalloped lateral band. **FOOD** Adult: Flowering plants (nectar). Larva: *Astragalus* spp., *Lotus* spp., wild buckwheat (*Eriogonum* spp.), knotweed (*Polygonum* spp.). **FOUND** Throughout the region, in various habitats from the coast to inland mountains. Can be lured into the garden if there are naturally occurring populations nearby.

Also known as *Plebejus acmon*. The larva may form a relationship with ants that is beneficial for both species: the ants obtain a nutrient from a special gland on the larva and, in turn, protect the larva from predators and parasites. See Glassberg (2001) for discussion of this and related species.

Incisalia augustinus

Incisalia augustinus

BROWN ELFIN

ADULT Below, forewings pale brown; hindwings brown, less commonly reddish brown, each with dark brown area. **WINGSPAN** 28 mm. **LARVA** Bright green; each segment with pair of creamy white and dark red dorsal bands. **FOOD** Adult: Flowering plants (nectar). Larva: Bearberry (*Arctostaphylos uva-ursi*) and other manzanitas (*Arctostaphylos* spp.), *Ceanothus velutinus* and other California-lilacs (*Ceanothus* spp.), as well as other plants. **FOUND** Throughout much of the U.S.

This species is often overlooked because of its plain brown color. Look for adults early in the spring.

Incisalia augustinus, larva

Incisalia eryphon

WESTERN PINE ELFIN

ADULT Below, forewings brown with few dark brown bands; hindwings light purplish brown with zigzag pattern (has been described as resembling a richly colored Turkish rug). **WINGSPAN** 32 mm. **FOOD** Adult: Flowering plants (nectar). Larva: Lodgepole pine (*Pinus contorta*) and other pines (*Pinus* spp.). **FOUND** Throughout the region.

Incisalia eryphon

Incisalia mossii

Incisalia mossii, larvae

Lycaena arota, male

Lycaena arota

Incisalia mossii

MOSS'S ELFIN

ADULT Below, forewings brown with dark brown postmedial lines edged in white; hindwings brown, two-toned, with dark brown postmedial lines edged in white. **WINGSPAN** 25 mm. **LARVA** Often red but less commonly green or yellow. **FOOD** Adult: Flowering plants (nectar). Larva: *Sedum spathulifolium* and other *Sedum* spp. **FOUND** Throughout the region, from the coast to the mountains, wherever the host is common. Adults stay close to the larval host plant.

Lycaena arota

TAILED COPPER

ADULT Above, male, wings coppery brown; female, wings light orange with brown spotting. Below, forewings mix of orange and gray with black spots; hindwings gray-brown with black spots and lines. Hindwing with tail. **WINGSPAN** 30 mm. **FOOD** Adult: Flowering plants (nectar). Larva: Currant or gooseberry (*Ribes* spp.) (leaves). **FOUND** Oregon and California.

The tailed hindwing makes this species one of the more easily identifiable coppers.

Lycaena cupreus, male and female, posed

Lycaena cupreus

LUSTROUS COPPER

ADULT Above, male, wings bright coppery orange with black spots and black border; female, wings similar to male's but yellowish orange. Below, forewings gray infused with orange, and with black spots; hindwings gray with black spots and orangish red line along outer margin. **WINGSPAN** 30 mm. **FOOD** Adult: Flowering plants (nectar). Larva: Dock (*Rumex* spp.). **FOUND** Throughout the region, most commonly in Oregon and California, at high elevations to above timberline.

Lycaena cupreus

Lycaena gorgon

GORGON COPPER

ADULT Above, male, wings coppery brown; female, wings dark brown with orangish yellow spots. Below, forewings light gray (may be infused with orange) with black spots; hindwings light gray with black spots and outer margin with row of black-capped orange crescents. **WINGSPAN** 32 mm. **LARVA** Pale green and covered with short white hairs. **FOOD** Adult: Flowering plants (nectar). Larva: *Eriogonum nudum* and other wild buckwheats (*Eriogonum* spp.). **FOUND** Oregon and California. On *E. nudum*, female deposits eggs on main flower stem, at the point where it branches into smaller flower stems.

Lycaena gorgon, female

Lycaena gorgon

Lycaena helloides

Lycaena xanthoides

Mitoura nelsoni

Lycaena helloides

PURPLISH COPPER

ADULT Above, male, wings brown with purplish iridescence and black spots; female, wings orange and brown with dark brown spots; both sexes with orange zigzag pattern along outer margin of hindwing. Below, forewings medium orange with black spots; hindwings pinkish to brownish gray with small black spots and reddish orange zigzag line along outer margin. **WINGSPAN** 32 mm. **FOOD** Adult: Flowering plants (nectar). Larva: Dock (*Rumex* spp.), knotweed (*Polygonum* spp.). **FOUND** Throughout the region, from the coast to the mountains.

Lycaena xanthoides

GREAT COPPER

ADULT Above, forewings brownish gray; hindwings brownish gray with narrow, orange band along outer margin. Below, forewings gray with black spots; hindwings gray with black spots and orange, scalloped band along outer margin. Hindwing with very short tail. **WINGSPAN** 44 mm. **FOOD** Adult: Flowering plants (nectar). Larva: Willow dock (*Rumex salicifolius*) and other docks (*Rumex* spp.). **FOUND** Oregon and California.

This species is the largest copper in the region.

Mitoura nelsoni

NELSON'S HAIRSTREAK

ADULT Above, wings brown. Below, forewings more brown than purple, each with ragged, dark brown line edged with white through middle; hindwings more purple than brown, each with ragged, dark brown line edged with white, often reduced, through middle. Hindwing with tail. **WINGSPAN** 25 mm. **FOOD** Adult: Flowering plants (nectar). Larva: Incense cedar (*Calocedrus decurrens*) and other cypresses (Cupressaceae). **FOUND** From British Columbia to California.

The adult is usually common every year in the spring. Butterfly experts disagree among themselves as to whether this species should stand alone or be incorporated into another species (see Glassberg [2001] for a discussion of hairstreaks).

Mitoura siva

Mitoura siva

JUNIPER HAIRSTREAK

ADULT Above, wings brown. Below, forewings more brown than purple, each with distinct, ragged, dark brown line edged with white through middle; hindwings more purple than brown, each with distinct, ragged, dark brown line edged with white through middle. Hindwing with tail. **WINGSPAN** 25 mm. **LARVA** Green with yellow to white lines; mimics host's foliage. **FOOD** Adult: Flowering plants (nectar). Larva: Juniper (*Juniperus* spp.). **FOUND** Throughout the region or only in California, depending upon which insect book one uses.

Butterfly experts disagree as to whether this species should stand alone or be incorporated into another species (see Glassberg [2001] for a discussion of hairstreaks).

Mitoura siva, larva

Satyrium californica

CALIFORNIA HAIRSTREAK

ADULT Below, forewings grayish brown with postmedial row of black spots; hindwings similar to forewings but also with row of orange marginal spots (often extending into forewing) and each with one orange-capped blue spot near tail. **WINGSPAN** 32 mm. **FOOD** Adult: Flowering plants (nectar). Larva: California-lilac (*Ceanothus* spp.), oak (*Quercus* spp.) and the antelope brush *Purshia tridentata*. **FOUND** Throughout the region.

Satyrium californica

Satyrium saepium

Satyrium saepium, larva

Satyrium sylvinus

Satyrium saepium

HEDGEROW HAIRSTREAK

ADULT Below, forewings gray-brown with distinct dark brown postmedial line edged in white; hindwings similar to forewings but each also with blue spot near tail. **WINGSPAN** 28 mm. **LARVA** Green with yellow dorsal and lateral stripes. **FOOD** Adult: Flowering plants (nectar). Larva: *Ceanothus cuneatus, C. velutinus*, and other California-lilacs (*Ceanothus* spp.). **FOUND** Throughout the region.

Satyrium sylvinus

SYLVAN HAIRSTREAK

ADULT Below, forewings gray to gray-brown with postmedial line of black spots; hindwings similar to forewings but each also with orange marginal spots and gray-blue spot near tail. **WINGSPAN** 32 mm. **LARVA** Green with distinct, yellow lateral line. **FOOD** Adult: Flowering plants (nectar). Larva: Willow (*Salix* spp.). **FOUND** Throughout the region. Can normally be found in same patch of willows year after year.

Satyrium tetra

MOUNTAIN MAHOGANY HAIRSTREAK

ADULT Below, forewings dark gray infused with brassy brown and each with faint, dark brown line edged with white through mid-

Satyrium tetra

Satyrium tetra, larva

dle; hindwings similar to forewings but each also with gray-blue spot near tail and sometimes with very faint row of light orange spots along outer margin. **WINGSPAN** 32 mm. **LARVA** Pale green with pale yellow lines and orangish red hairs on anterior end. **FOOD** Adult: Flowering plants (nectar). Larva: *Cercocarpus betuloides* and other mountain-mahoganies (*Cercocarpus* spp.). **FOUND** Oregon and California.

Strymon melinus

GRAY HAIRSTREAK

ADULT Above, forewings gray; hindwings gray, each with one orangish red spot near tails. Below, wings gray with postmedial line of orangish red spots edged with black and white. **WINGSPAN** 32 mm. **LARVA** Variable—green, yellow, or red (shown). **FOOD** Adult: Flowering plants (nectar). Larva: Mallow family (Malvaceae) and many other plants (flowers and seed pods). **FOUND** Throughout the U.S.

Strymon melinus

Strymon melinus, larva

TUSSOCK MOTHS Family Lymantriidae

This is a somewhat large family of moths, although relatively few species are native to the United States and Canada. Similar to the noctuid moths (Noctuidae), the adults are medium in size, stout, and nondescript in color, usually brown to grayish white. The females of some species are wingless. The males have feathery antennae. The adults of all species are short-lived and do not feed. The slender larvae typically have long tufts of hair, called *tussocks,* at each end of the body (hence the common name for the family) and shorter, often brightly colored tufts on the back. The larvae of most species commonly feed on the foliage of trees and shrubs and are often seen in gardens. The pupae are enclosed in loose cocoons that often incorporate larval hairs. Some species are serious pests of forest and shade trees.

Leucoma salicis, adult and pupal skin

Leucoma salicis, larva

Leucoma salicis

SATIN MOTH

ADULT Above, forewings satiny or shiny white with leading margin edged with light yellow. Legs banded black and white. **WINGSPAN** 50 mm. **LARVA** With series of large, creamy white dorsal spots in between which are pairs of red bumps; sides covered with white hairs. **FOOD** Larva: Willow (*Salix* spp.); black cottonwood (*Populus balsamifera*), quaking aspen (*P. tremuloides*), and other poplars (*Populus* spp.). **FOUND** Very common throughout the U.S.

This species was introduced from Europe. Pete has seen huge poplars that have been completely stripped of their leaves by the larvae. When threatened, the adult releases from the base of its head a liquid that repels predators (see photograph).

Orgyia antiqua

RUSTY TUSSOCK MOTH

ADULT Male, above, forewings brown, each with white spot; antennae large. Female, covered with gray hairs; with vestigial wings. **WINGSPAN** 30 mm. **LARVA** Black; covered with white hairs arising from red bumps; two tussocks of long, black hairs near head and one tussock of long black hairs at posterior end of body; four gold dorsal tufts. **FOOD** Larva: Willow (*Salix* spp.), red alder (*Alnus rubra*) and other alders (*Alnus* spp.), as well as many other trees and shrubs. **FOUND** Throughout the U.S.; very common in coastal areas.

After the female emerges from her cocoon, she clings to its side and releases a pheromone to attract a male. Detecting the pheromone with his antennae, the male locates the female and mates with her. The female then lays her eggs on her cocoon. The eggs overwinter on the cocoon, and the larvae emerge in late spring.

Orgyia antiqua, male

Orgyia antiqua, females with eggs

Orgyia antiqua, larva

Orgyia pseudotsugata

DOUGLAS-FIR TUSSOCK MOTH

ADULT Male, above, forewings mottled with black, brown, and gray, and each with white spot; antennae large. Female, covered with gray hairs; with vestigial wings. **WINGSPAN** 33 mm. **LARVA** Black; covered with white hairs arising from red bumps; two tussocks of long black hairs near head and one tussock of long black hairs at posterior end of body; four pale white dorsal tufts with brown tips. **FOOD** Larva: White fir (*Abies concolor*), grand fir (*A. grandis*), Douglas-fir (*Pseudotsuga menziesii*). **FOUND** Throughout the region.

After the female emerges from her cocoon, she clings to its side and releases a pheromone to attract a male. Detecting the pheromone with his antennae, the male lo-

Orgyia pseudotsugata, male

Orgyia pseudotsugata, larvae

cates the female and mates with her. The female then lays her eggs on the old cocoon and covers them with her body hair, which camouflages and physically protects the eggs from predators. The eggs overwinter on her cocoon, and the larvae emerge in late spring.

NOCTUID MOTHS Family Noctuidae

This is the largest family of lepidopterans and consists of approximately one-third of all described moths in North America. The adults vary greatly in size and color, but most are medium in size, stout-bodied, and a dull brown to gray in color with complex patterns of subtle lines and spots on the wings. In most species, the forewings are somewhat narrow, and the hindwings are broad. When at rest, most species hold their wings rooflike over the body. The adults of most species are nocturnal; some are brightly colored day-flyers. The majority of moths that are found at lights at night belong to this family. The thick-bodied larvae, many of which are called *cutworms*, are usually dull in color like the adults. Most species are smooth-skinned and either hairless or only sparsely hairy, although others are covered with hair. Most feed on a wide range of plants. The larvae of a number of species, including cutworms, are serious agricultural and garden pests.

Acronicta hesperida

Acronicta hesperida

ADULT Above, forewings gray with distinct, small, black spots along outer margin. **WINGSPAN** 53 mm. **LARVA** Black; with dorsal bands of brownish orange hairs, lateral bands of light brown hairs, and few, scattered, small

Acronicta hesperida, larva

tufts of black hairs; head black. **FOOD** Larva: Red alder (*Alnus rubra*) and other alders (*Alnus* spp.). **FOUND** Throughout the region. Larvae found in the fall on trunks of host trees or crossing roads looking for places to pupate.

Acronicta lepusculina

Acronicta lepusculina

COTTONWOOD DAGGER MOTH

ADULT Above, forewings gray with faint, black streaking. **WINGSPAN** 50 mm. **LARVA** Green and covered with long, recurved, golden yellow hairs (when fully grown, larva is very conspicuous on tree foliage). **FOOD** Larva: Scouler's willow (*Salix scouleriana*) and other willows (*Salix* spp.), poplar (*Populus* spp.). **FOUND** Throughout the U.S.

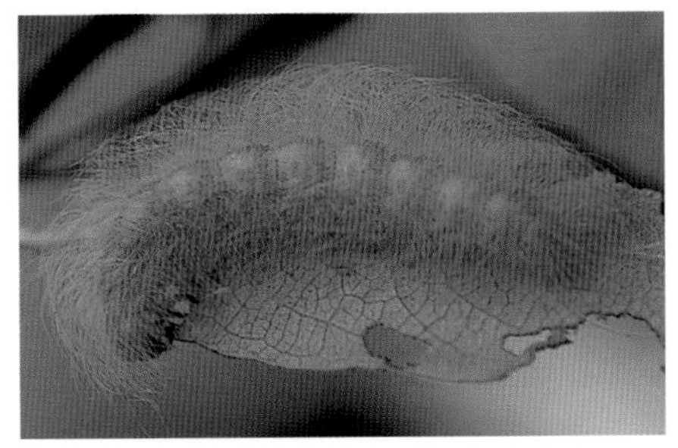

Acronicta lepusculina, larva

Acronicta perdita

ADULT Above, forewings gray suffused with black; hindwings white in male, gray in female. **WINGSPAN** 40 mm. **LARVA** Dark brown with white lateral band and tufts of reddish brown to black hairs emerging from reddish brown bumps. **FOOD** Larva: California-lilac (*Ceanothus* spp.), manzanita (*Arctostaphylos* spp.), and probably other shrub species. **FOUND** Throughout the region. Look for the larva in July.

Acronicta perdita

Acronicta perdita, larva

Agrotis ipsilon

Agrotis ipsilon

BLACK CUTWORM

ADULT Above, forewings light brown, each with dark brown medial area and small, black postmedial wedge. **WINGSPAN** 45 mm. **LARVA** Brown to black, often with a greasy or shiny look. **FOOD** Larva: Herbaceous plants. **FOUND** Throughout the U.S., commonly in gardens.

This species has several generations per year.

Alypia ridingsii

Alypia ridingsii

RIDINGS' FORESTER

ADULT Above, forewings black, each with two large white patches (the cells within the patches have black borders), one small white patch between the two large patches; white shoulder patches. First and second pair of legs with bright orange, hairy patches. **WINGSPAN** 34 mm. **LARVA** White with black spots and lines, pale orange bands, and sparsely covered with long, white hairs. **FOOD** Larva: Fireweed (*Epilobium angustifolium*) and other related plants. **FOUND** Throughout the region.

Alypia ridingsii, larva

Amphipyra pyramidoides

COPPER UNDERWING

ADULT Above, forewings mottled brown with light brown submarginal border; hindwings orange. **WINGSPAN** 50 mm. **LARVA** Green

Amphipyra pyramidoides

Amphipyra pyramidoides, larva

with yellow lateral line and prominent hump on posterior end. **FOOD** Larva: Bigleaf maple (*Acer macrophyllum*), deciduous oaks (*Quercus* spp.), and other broadleaf trees and shrubs. **FOUND** Throughout the region. Larva common on bigleaf maple and deciduous oaks.

This species is not a true underwing (*Catocala* spp.).

Aseptis binotata, larva

Aseptis binotata

ADULT Above, forewings dark brown, each with pale yellow postmedial spot. **WINGSPAN** 35 mm. **LARVA** Green with white speckling and red-and-white lateral line. **FOOD** Larva: Currant or gooseberry (*Ribes* spp.) and other broadleaf shrubs. **FOUND** Throughout the region. Larva is common on gooseberry.

Autographa biloba

Autographa biloba

BILOBED SEMILOOPER

ADULT Above, forewings mix of light and dark brown, each with light brown area along outer margin, and distinct, bilobed, silvery white spot in middle. **WINGSPAN** 22 mm. **LARVA** Green; head with black lateral stripe. **FOOD** Larva: Herbaceous plants, including garden plants. **FOUND** Throughout the U.S. Adult is attracted to lights at night. Larva is common in gardens.

Also known as *Megalographa biloba*.

Autographa californica

Autographa californica

ALFALFA LOOPER

ADULT Above, forewings mottled gray and black, each with Y-shaped, silvery white spot in middle. **WINGSPAN** 40 mm. **LARVA** Green with white lateral line. **FOOD** Larva: Mainly herbaceous plants. **FOUND** Very common throughout the region. Larva has been known to sample garden plants.

Despite its common name, this species has a wide range of host plants.

Autographa corusca

Behrensia conchiformis

Catocala aholibah

Autographa corusca

ADULT Above, forewings mottled purple-brown and pink, each with medially constricted, silvery white spot in middle. **WINGSPAN** 40 mm. **FOOD** Larva: Alder (*Alnus* spp.) and probably other plants. **FOUND** Throughout the region. Adult is attracted to lights at night.

Behrensia conchiformis

ADULT Above, forewings mottled gray with green highlights and each with wide, dark brown, medial band containing large, white spot. **WINGSPAN** 25 mm. **FOOD** Larva: Plants—specific host unknown. **FOUND** Throughout the region. Adult is attracted to lights at night in the spring.

Catocala aholibah

AHOLIBAH CATOCALA

ADULT Above, forewings mottled gray and brown with black lines, and each with small, white spot ringed with black in center; hindwings pinkish red with narrow black medial band and broad black marginal band. **WINGSPAN** 79 mm. **LARVA** Grayish brown, sometimes with pink tint; two dorsal bumps; first three pairs of legs (true legs) pinkish red. **FOOD** Larva: Oregon oak (*Quercus garryana*), California black oak (*Q. kelloggii*), and other oaks (*Quercus* spp.). **FOUND** Throughout the

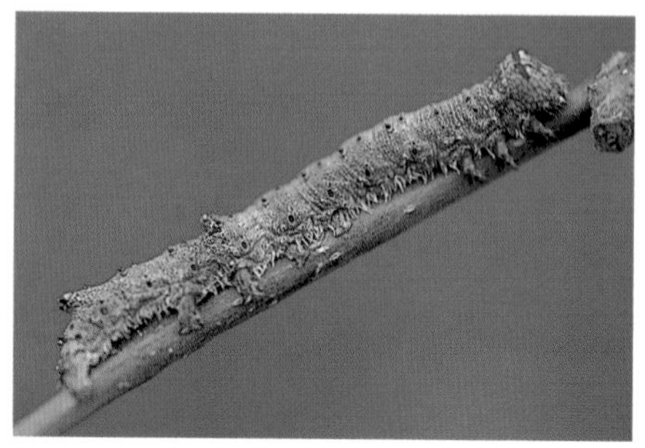
Catocala aholibah, larva

Catocala aholibah, posed adult with hindwings exposed

region. Look for larvae on oak stems in early spring (note how well they are camouflaged).

Cerastis enigmatica

Cerastis enigmatica

ENIGMATIC MOTH

ADULT Above, forewings mottled dark reddish brown with pale brown outer margin and each with wide, roughly V-shaped spot in middle. **WINGSPAN** 30 mm. **FOOD** Larva: Salmonberry (*Rubus spectabilis*). **FOUND** Throughout the region.

Cosmia calami

Cosmia calami

AMERICAN DUN-BAR MOTH

ADULT Above, forewings light brown, each with wide, darker brown medial band containing two brown spots outlined in pale white. **WINGSPAN** 32 mm. **FOOD** Larva: Oregon oak (*Quercus garryana*), California black oak (*Q. kelloggii*), and other oaks (*Quercus* spp.). **FOUND** Throughout the region.

Dargida procinctus

Dargida procinctus

OLIVE-GREEN CUTWORM

ADULT Above, forewings brown to purplish brown with dark brown and light brown crosshatching. **WINGSPAN** 45 mm. **LARVA** Green with white and red bands. **FOOD** Larva: Grasses. **FOUND** Throughout the region. Adult is common around lights at night.

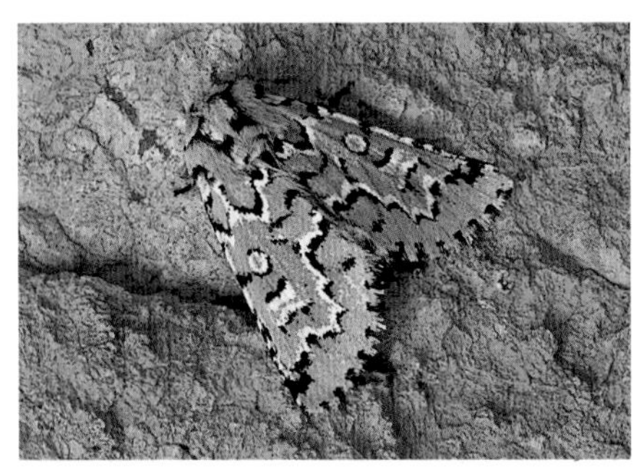
Feralia februalis

Feralia februalis

ADULT Above, forewings pale green with wavy, white-edged black lines. **WINGSPAN** 35 mm. **LARVA** Green with sparse, small, pale white spots, lateral white line often bordered with red, and large dorsal hump. **FOOD** Larva: Oak (*Quercus* spp.) and possibly other plants. **FOUND** Throughout the region.

The adult is active late winter and early spring.

Leucania farcta

Leucania farcta

ADULT Above, forewings brownish pink, each with thin, clear white line in middle. **WINGSPAN** 35 mm. **FOOD** Larva: Grasses. **FOUND** Throughout the region.

The adults fly in late summer.

Orthosia hibisci

ADULT Above, forewings reddish brown and gray with lighter gray to brown outer margin and each with two large, grayish brown spots outlined in pale white in middle. **WINGSPAN** 35 mm. **LARVA** Green with white speckling

Orthosia hibisci

Orthosia hibisci, larva

and pale white lines. **FOOD** Larva: Woody trees and shrubs. **FOUND** Throughout the region.

The larvae are common in spring and early summer.

Papaipema sauzalitae

Papaipema sauzalitae

ADULT Above, forewings mostly reddish brown, each with dark brown submarginal band and yellow spots in middle. **WINGSPAN** 40 mm. **LARVA** Pale white with sparse hairs; head dark brown. **FOOD** Larva: Himalayan blackberry (*Rubus discolor*) and other *Rubus* spp. (canes). **FOUND** Range unknown, but this species probably occurs throughout the region. Frass (larval fecal material) near a hole on the cane (see photograph) indicates larval activity (all *Papaipema* spp. are stem borers).

Papaipema sauzalitae, exit hole and frass on cane

Pseudaletia unipuncta

ARMYWORM MOTH

ADULT Above, forewings pale brown, each with tiny, white medial spot and fine, black line extending inward from tip. **WINGSPAN** 45 mm. **FOOD** Larva: Herbaceous plants. **FOUND** Throughout the U.S., commonly in yards and gardens (mowing a lawn will often flush out adults). Adult is attracted to lights at night.

Pseudaletia unipuncta

Scoliopteryx libatrix

Syngrapha alias

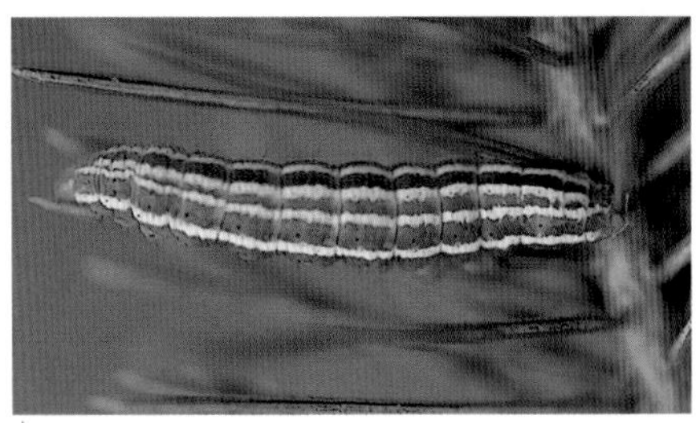

Syngrapha alias, larva

Xylena nupera

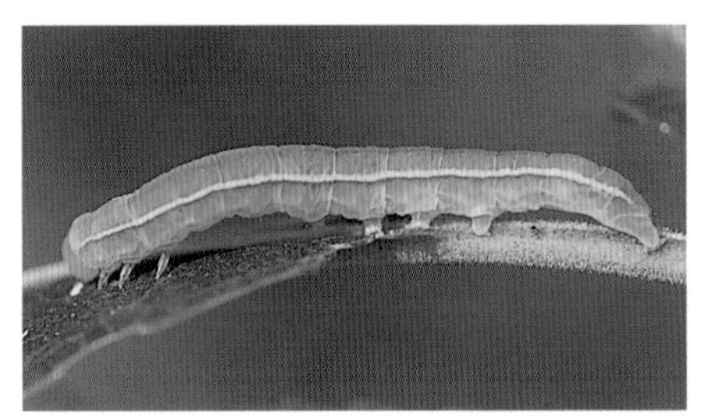

Scoliopteryx libatrix, larva

Scoliopteryx libatrix

THE HERALD

ADULT Above, forewings reddish brown, each with two small but distinct white spots and two gray medial lines. **WINGSPAN** 45 mm. **LARVA** Long; translucent green with light yellow lateral line. **FOOD** Larva: Willow (*Salix* spp.), black cottonwood (*Populus balsamifera*) and other poplars (*Populus* spp.). **FOUND** Throughout the U.S.

The feeding larvae silk together the young leaves of the host plant.

Syngrapha alias

ADULT Above, forewings mottled gray and dark brown, each with bifurcated, white, medial spot. **WINGSPAN** 16 mm. **LARVA** Green with white, longitudinal lines. **FOOD** Larva: Sitka spruce (*Picea sitchensis*) and other spruces (*Picea* spp.). **FOUND** Throughout the region.

Xylena nupera

ADULT At rest (with wings folded close to body), resembles a twig. Above, forewings elongated, light brown, each with black medial streak and black medial spot. **WINGSPAN** 50 mm. **FOOD** Unknown. **FOUND** Throughout the region.

Zale lunata

Zale lunata

MOON UMBER MOTH

ADULT Above, forewings mottled brown with wavy, black and dark brown lines and often with pale white marginal patches. **WINGSPAN** 48 mm. **LARVA** Light brown to gray with two swollen areas dorsally, one near middle of body and one near posterior end. **FOOD** Larva: Coastal willow (*Salix hookeriana*), shining willow (*S. lucida*), and other willows (*Salix* spp.), as well as other broadleaf plants. **FOUND** Throughout the region.

The adult is very common in the spring and late summer.

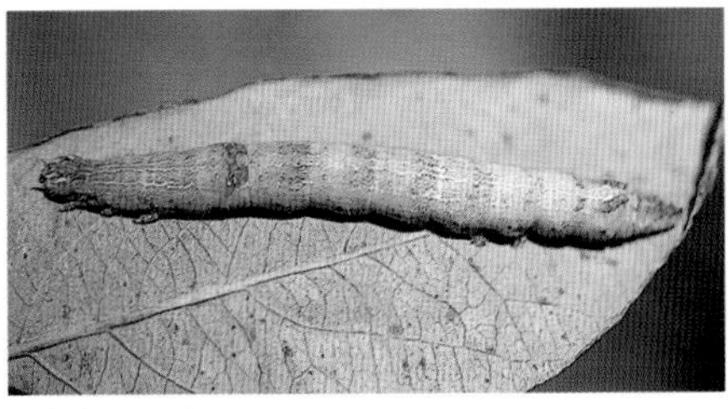

Zale lunata, larva

Zotheca tranquilla

TRANQUIL MOTH

ADULT Above, forewings light green, each with green medial band and two thin, wavy, green lines. **WINGSPAN** 35 mm. **LARVA** With dorsal row of large yellow spots surrounded with black; sides yellow with small, black spots. **FOOD** Larva: Blue elderberry (*Sambucus mexicana*) and other elderberries (*Sambucus* spp.). **FOUND** Throughout the region. Look for larva in June and July on partially defoliated host plant; larva is usually inside a folded leaf silked together.

Zotheca tranquilla

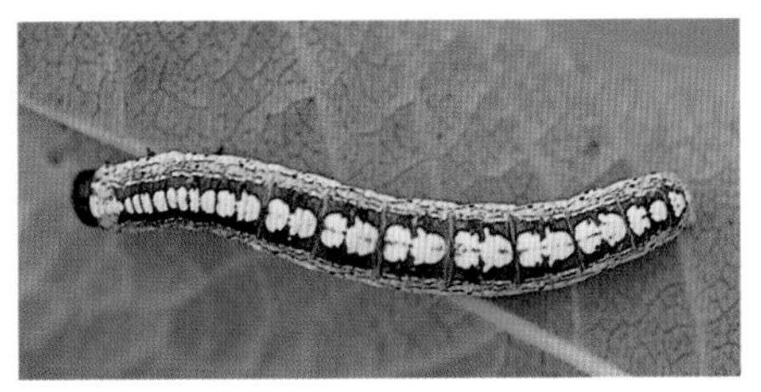

Zotheca tranquilla, larva

PROMINENT MOTHS Family Notodontidae

The adults in this family of common moths are similar to the noctuid moths (Noctuidae) in general appearance—they are medium-sized, stout-bodied, and usually a drab brown to gray. Also like noctuids, they are abundant at lights at night. Unlike many noctuids, the males have comblike antennae. The family name (Greek, *not*, back; *odont*, tooth) refers to the fact that, in some species, the forewings have a backward-projecting, toothlike tuft on the posterior margin, which shows "prominently" when the wings are held rooflike over the body at rest. The larvae are variously patterned and shaped, often mottled or striped with conspicuous bumps. When threatened, the larvae of some species lift the anterior and posterior ends of the body and remain motionless. Notodontid larvae feed on the foliage of a large variety of trees and shrubs, and some species feed in groups. They overwinter as mature larvae or pupae usually in cells in the soil or in loose cocoons in the ground. Some species are orchard pests.

Clostera brucei

Clostera brucei, larva

Clostera brucei

ADULT Above, forewings gray and pale brown, more reddish brown toward outer margin, and with pale white lines. Abdomen with posterior brush. **WINGSPAN** 30 mm. **LARVA** Finely mottled gray and black, covered with fine white hairs, and with orange spots. **FOOD** Larva: Coastal willow (*Salix hookeriana*), shining willow (*S. lucida*), and other willows (*Salix* spp.), and possibly other broadleaf plants. **FOUND** Throughout the region. Larva found inside one or more leaves which it has silked together.

The larvae are common in the fall, less common in the summer.

Furcula scolopendrina

WHITE FURCULA

ADULT Above, forewings white with black bands flecked with brown and small, black spots along outer margin. Body and legs covered with long, silky, white hairs. **WINGSPAN** 45 mm. **LARVA** In later instars, green with small, green spots ringed with red, and reddish brown to dark brown saddle; prolegs very long, thin, and, when at rest, held straight out horizontally from body. **FOOD** Larva: Coastal willow (*Salix hookeriana*), shining willow (*S. lucida*), and other willows (*Salix* spp.); poplar (*Populus* spp.). **FOUND** Common throughout the region. Adults attracted to lights at night.

The British call these moths *kitten sallows* (sallow is a common name for a willow in Britain) probably because the downy hair of the adult is like a kitten's fur. The adult lays a single, flattish, black egg on a leaf of the host tree. The pupating larva spins a silken cocoon with bits of the substrate incorporated into the cocoon so that the cocoon closely matches the substrate. When a larva is threatened, it lifts the front end of its body off the leaf and waves its elongated pair of legs.

Furcula scolopendrina

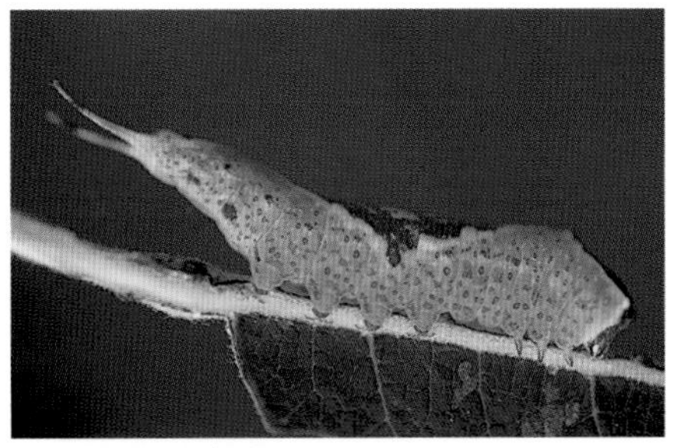

Furcula scolopendrina, larva

Nadata oregonensis

OREGON PROMINENT

ADULT Above, forewings pale brown, each with two reddish brown medial lines and small, white medial spot bordered with reddish brown. Very similar to *Nadata gibbosa* (not described) but forewings of latter yellowish brown. **WINGSPAN** 55 mm. **LARVA** Green with small, pale yellow spots and white lateral line. **FOOD** Larva: Oregon oak (*Quercus garryana*), California black oak (*Q. kelloggii*), and other oaks (*Quercus* spp.). **FOUND** Common throughout the region.

Nadata oregonensis

Nadata oregonensis, larva

Oligocentria pallida

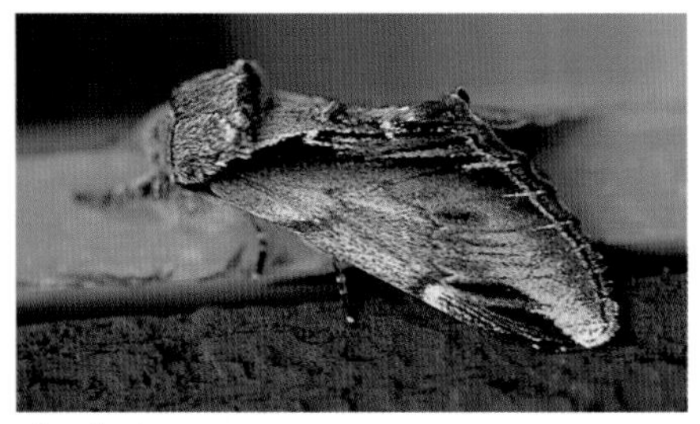

Pheosia rimosa

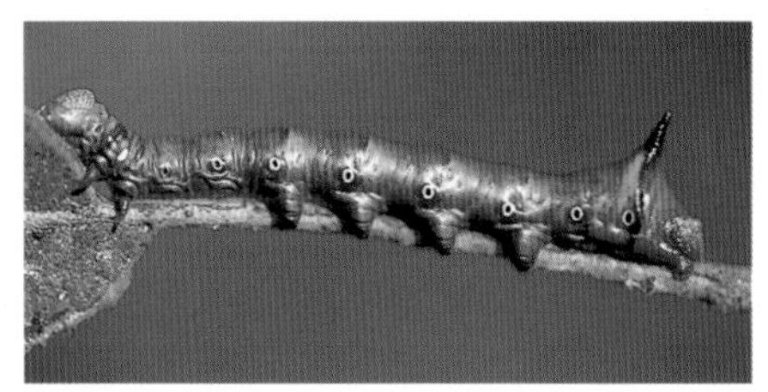

Pheosia rimosa, larva

Schizura concinna

Oligocentria pallida

PALLID PROMINENT

ADULT Resembles a twig (often rests as shown in photograph). Above, forewings light yellow with splattering of reddish brown, and gray leading margin. **WINGSPAN** 45 mm. **LARVA** First two body segments green with dark brown dorsal band, other segments reticulated with reddish brown (this pattern makes the larva difficult to detect on a leaf) and with yellow, Y-shaped dorsal mark on posterior end. **FOOD** Larva: Willow (*Salix* spp.), cascara (*Rhamnus purshiana*), and many other woody trees and shrubs. **FOUND** Throughout the region.

Pheosia rimosa

BLACK-RIMMED PROMINENT

ADULT Above, forewings light gray and brown with dark brown to black inner margin and black postmedial dashes. **WINGSPAN** 60 mm. **LARVA** Large; variably colored—usually shiny brown with prominent lateral spiracles and tail horn. **FOOD** Larva: Willow (*Salix* spp.), poplar (*Populus* spp.). **FOUND** Very common throughout the region. Larva is easy to find in the fall feeding on leaves.

Schizura concinna

RED-HUMPED CATERPILLAR

ADULT Above, forewings gray, each with reddish brown band across inner margin. **WING-**

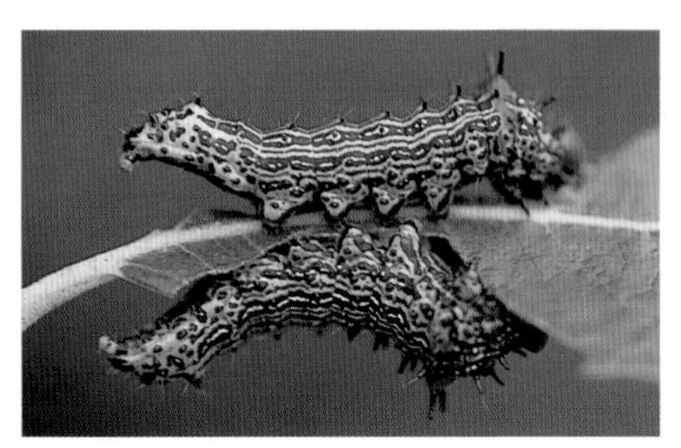

Schizura concinna, larvae

SPAN 35 mm. **LARVA** Striped with yellow, black, and white and with two red dorsal bumps; head red. **FOOD** Larva: Trees and shrubs such as bigleaf maple (*Acer macrophyllum*), the California-lilac *Ceanothus velutinus*, *Rosa* spp., and willow (*Salix* spp.). **FOUND** Throughout the U.S.

The larvae are common from late summer to early fall; they are gregarious feeders and very conspicuous. When threatened, the larvae release an odor that is probably a repellent.

BRUSH-FOOTED BUTTERFLIES Family Nymphalidae

This is the largest and most diverse family of butterflies in the world. Their distribution is worldwide, and many species are common; these are the butterflies most frequently seen in gardens. Although they come in a wide spectrum of colors and shapes, they are generally medium in size and have broad wings. The one characteristic they all share is the forelegs, which are greatly reduced and useless for walking. The forelegs are covered with hair, somewhat resembling a brush, giving rise to the common name of the family. Although their flight habits and resting positions vary, most species are strong flyers. The adults usually feed on nectar, sap, feces, or carrion. Most overwinter as larvae; some overwinter as adults, generally unusual for butterflies. The larvae greatly vary in shape but usually are cylindrical and covered with spines. They feed on a great diversity of flowering plants. The pupae, which hang by the tip of the abdomen, also greatly vary in shape but usually have prominent tubercles.

Adelpha bredowii

CALIFORNIA SISTER

ADULT Above, forewings dark brown, each with medial band of large white spots and large, round, orange patch at tip; hindwings similar to forewings but without orange patch. Below, forewings each with reddish brown, pale blue, and white bands and large, round, brownish orange patch at tip; hindwings similar to forewings but without orange patch. Similar to Lorquin's admiral (*Limenitis lorquini*); forewings of latter, above, with orange patch extending to outer margin. **WINGSPAN** 86 mm. **LARVA** Green with four to six pairs of light brown dorsal spines. **FOOD** Adult: Flowering plants (nectar) and mammal feces. Larva: Tan(bark) oak (*Lithocarpus*

Adelpha bredowii

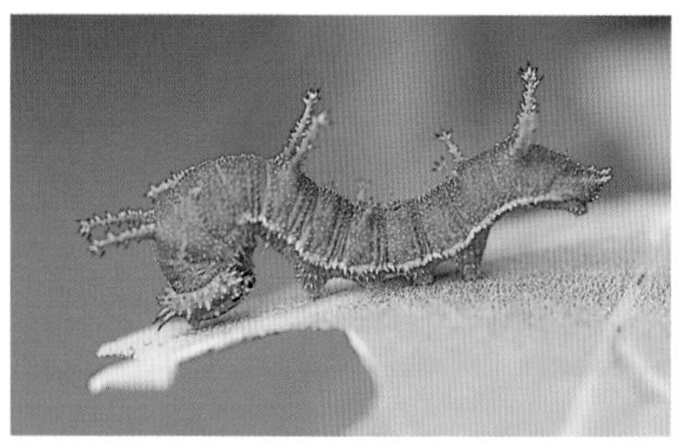

Adelpha bredowii, larva

Boloria epithore

Cercyonis pegala

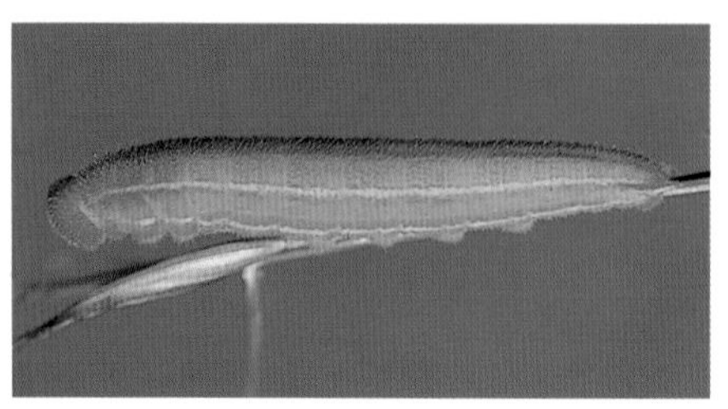

Cercyonis pegala, larva

densiflorus), oak (*Quercus* spp.). **FOUND** From Washington to California.

Boloria epithore

WESTERN MEADOW FRITILLARY

ADULT Above, wings orange with black spots and lines. Below, forewings paler orange with black spots and lines; hindwings purplish brown becoming orangish brown at base and with medial band of pale yellow spots. **WINGSPAN** 40 mm. **FOOD** Adult: Flowering plants (nectar). Larva: Violet (*Viola* spp.). **FOUND** Throughout the region.

Also known as *Clossiana epithore*. This is the smallest fritillary in the region and usually the first fritillary to fly in the spring.

Cercyonis pegala

COMMON WOOD NYMPH

ADULT Quite variable in color and pattern. Above, wings light to dark brown. Below, forewings brown with dark brown striations and each with two large eyespots, often ringed with yellow or enclosed in large yellow patch; hindwings brown with dark brown striations and with or without small submarginal eyespots. Female large. **WINGSPAN** 73 mm. **LARVA** Green with light yellow lateral line and pair of pink tails. **FOOD** Adult: Flowering plants (nectar). Larva: Grasses. **FOUND** Throughout most of the U.S. Adult often emerges from inland grasslands in large numbers in the summer; female can be found into late summer.

The adult females are long-lived.

Chlosyne palla

NORTHERN CHECKERSPOT

ADULT Above, wings reddish orange with bands of paler orange and yellow spots edged in black. Below, hindwings with alternating bands of orangish red and creamy white outlined in black. Similar to chalcedona checkerspot (*Euphydryas chalcedona*) but smaller and not as common. **WINGSPAN** 40 mm. **FOOD** Larva: *Aster oregonensis* and other asters (*Aster* spp.). **FOUND** Throughout the region.

Chlosyne palla

Coenonympha tullia

CALIFORNIA RINGLET

ADULT Above, wings grayish white. Below, forewings reddish brown, two-toned with basal half darker than marginal half, and usually with small eyespot(s) (or none); hindwings similar to forewings but grayish brown. **WINGSPAN** 44 mm. **LARVA** Green to brownish with pair of short, pink tails. **FOOD** Larva: Grasses. **FOUND** Throughout the region. Adult is common in tall grass.

The adult is often mistaken for a moth because of its size, color, and habit of flying low to the ground (typical behavior of a butterfly species whose larvae eat grasses). It is difficult to locate because it often flies erratically, and when it lands, closes its wings and lies flat against the vegetation.

Chlosyne palla

Coenonympha tullia

Danaus plexippus

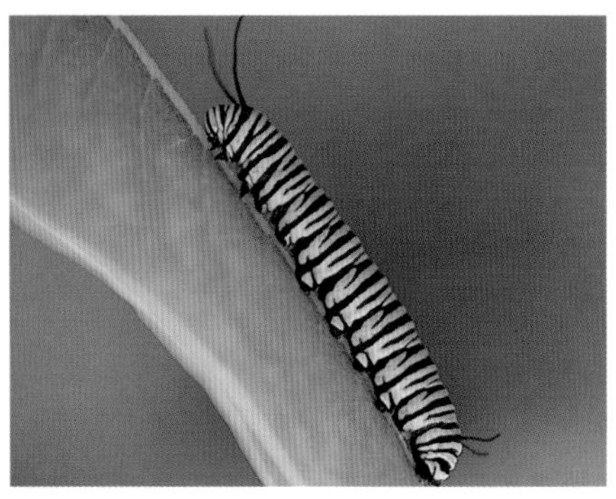

Danaus plexippus, larva

Danaus plexippus

MONARCH

ADULT Above, wings deep orange with wide, black veins and black margins with white spots. Below, wings similar to above but paler orange. **WINGSPAN** 102 mm. **LARVA** Banded white, yellow, and black with pair of long, black appendages at anterior and posterior ends. **FOOD** Adult: Flowering plants (nectar). Larva: Showy milkweed (*Asclepias speciosa*) and other milkweeds (*Asclepias* spp.). **FOUND** Throughout most of North America.

Euphydryas chalcedona

CHALCEDONA CHECKERSPOT

ADULT Above, wings black with white spots throughout and orangish red spots mainly on outer margin (coastal populations tend to be very dark, whereas inland populations are often much more brightly colored). Below, forewings reddish orange with row(s) of white spots; hindwings with alternating bands of orangish red and creamy white outlined in black. Similar to northern checkerspot (*Chlosyne palla*) but larger and more common. **WINGSPAN** 50 mm. **LARVA** Black with

Euphydryas chalcedona

bristly spines and orange spots. **FOOD** Adult: Flowering plants (nectar). Larva: English plantain (*Plantago lanceolata*), the monkey-flower *Mimulus aurantiacus, Keckiella lemmonii,* California figwort (*Scrophularia californica*), and many other plants. **FOUND** From British Columbia to California.

This species is difficult to separate from the colon, Edith's, and anicia checkerspots, which are not included in this field guide; in fact, lepidopterists disagree on the number of species in this group (see Glassberg [2001] and Scott [1986]).

Euphydryas chalcedona

Junonia coenia

BUCKEYE

ADULT Above, forewings dark brown, each with two orange bars and large eyespot; hindwings dark brown, each with two eyespots, one much larger than the other. **WINGSPAN** 60 mm. **LARVA** Grayish black with yellowish orange spots and bristly, iridescent dark blue spines. **FOOD** Adult: Flowering plants (nec-

Euphydryas chalcedona, larvae

Junonia coenia

Junonia coenia, larva

Limenitis lorquini

tar). Larva: Figwort family (Scrophulariaceae) and other herbaceous plants. **FOUND** Oregon and California; common in gardens, especially those that abound in the weeds Persian speedwell (*Veronica persica*) and *V. serpyllifolia*.

Limenitis lorquini

LORQUIN'S ADMIRAL

ADULT Above, forewings dark brown, each with medial band of large white spots and elongated, orange patch at tip; hindwings similar to forewings but without orange patch. Below, wings reddish brown with bands of white spots bordered with black. Similar to California sister (*Adelpha bredowii*); forewings of latter, above, with orange patch not extending to outer margin. **WINGSPAN** 70 mm. **LARVA** Bird-dropping mimic—mottled dark brown and grayish white with splash of grayish white dorsally; two clublike appendages posterior to horned head. **FOOD**

Limenitis lorquini

Adult: Flowering plants (nectar). Larva: Coastal willow (*Salix hookeriana*), shining willow (*S. lucida*), and other willows (*Salix* spp.); black cottonwood (*Populus balsamifera*); apple (*Malus* spp.); plums and cherries (*Prunus* spp.). **FOUND** Throughout the region.

Also known as *Basilarchia lorquini*. The early instar larva overwinters in a larva-modified leaf called a *hibernaculum* (see photograph). The leaf stem is silked to the twig of a tree so that the hibernaculum remains on the tree all winter. On sunny days in winter, we have seen larvae sunning themselves outside their hibernacula.

Limenitis lorquini, larva

Nymphalis antiopa

MOURNING CLOAK

ADULT Above, wings velvety, dark brownish maroon with creamy yellow marginal band, next to which is row of blue spots. Below, wings dark brown with pale white marginal band. Wings with ragged outer margin—

Limenitis lorquini, hibernaculum and larva

Nymphalis antiopa

Nymphalis antiopa

Nymphalis antiopa, larva

when closed, butterfly resembles a dead leaf, a common trait in butterflies that overwinter as adults. **WINGSPAN** 85 mm. **LARVA** Velvety black with white speckling, rows of bristly spines, and longitudinal row of orangish red dorsal spots. **FOOD** Adult: Flowering plants (nectar). Larva: Coastal willow (*Salix hookeriana*) and other willows (*Salix* spp.), poplar (*Populus* spp.), and many other woody plants. **FOUND** Throughout the U.S. Look below defoliated areas of host plant to find larvae, which feed in groups.

Adult may often be seen sunning itself in warm weather, even in the middle of winter.

Nymphalis californica

CALIFORNIA TORTOISESHELL

ADULT Above, forewings rich, golden orange with black spots and black margins. Hindwings similar to forewings but with fewer to no black spots. Below, wings mottled gray-brown, two-toned with basal half darker than

Nymphalis californica

marginal half. **WINGSPAN** 60 mm. **LARVA** Black with white to pale yellow speckling with yellow speckling concentrated at black dorsal stripe, and with bristly spines. **FOOD** Adult: Flowering plants (nectar). Larva: *Ceanothus velutinus*, deer brush (*C. integerrimus*), mountain whitethorn (*C. cordulatus*), and other California-lilacs (*Ceanothus* spp.). **FOUND** Throughout the region. Not common on the coast; prefers inland mountains.

Every few years this species' populations build up to very high numbers, with individuals emigrating over large areas. This increase is followed by a dramatic drop in numbers. The year 2005 in northern California saw such a dramatic increase in population size.

Nymphalis californica, larva

Nymphalis milberti

MILBERT'S TORTOISESHELL

ADULT Above, forewings each with basal half dark brown, with two orange spots, marginal

Nymphalis milberti

Nymphalis milberti

Oeneis nevadensis

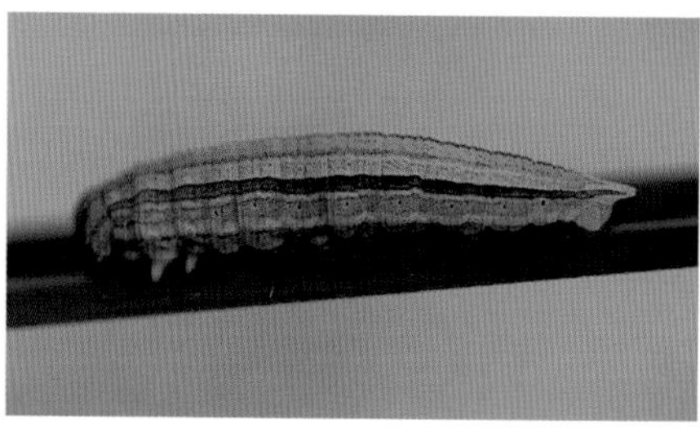
Oeneis nevadensis, larva

Phyciodes mylitta

half banded with pale yellow and orange and bordered dark brown; hindwings similar to forewings but dark brown outer margin with blue spots and leading margin without orange spots. Below, wings distinctly two-toned—basal half dark brown, marginal half light brown. **WINGSPAN** 50 mm. **LARVA** Black with yellow dorsal band, light green lateral lines, and bristly spines. **FOOD** Adult: Flowering plants (nectar). Larva: Stinging nettle (*Urtica* spp.). **FOUND** Throughout the region.

Oeneis nevadensis

GREAT ARCTIC

ADULT Above, wings pale brown to orangish brown. Below, forewings orangish brown with gray and brown mottling on tip and each with black eyespot with white center; hindwings mottled with brown and gray. **WINGSPAN** 60 mm. **LARVA** Light brown with dark brown dorsal lines and dark brown lateral band; body tapers posteriorly to pair of small tails. **FOOD** Larva: Grasses and grass-like plants. **FOUND** Throughout the region at higher elevations.

This is the largest and most widespread western arctic in the region. It has a two-year life cycle with the adults more common in even years.

Phyciodes mylitta

MYLITTA CRESCENT

ADULT Above, wings bright orange with black lines and bands. Below, wings variegated and much paler orange than above. Similar in size to and often flies with the field crescent (*Phyciodes pulchellus*) but latter is much darker above. **WINGSPAN** 14 mm. **LARVA** Brown with creamy white spots or lines; spiny. **FOOD** Adult: Flowering plants (nectar). Larva: Thistle (*Cirsium* spp.). **FOUND** Throughout the region.

The introduction of weedy thistles from Europe has probably helped this species increase its numbers and range.

Phyciodes pulchellus

Phyciodes pulchellus

FIELD CRESCENT

ADULT Above, wings dark brown with rows of orange or light yellow patches and spots. Below, wings variegated with different shades of pale orange. Similar in size to and often flies with the mylitta crescent (*Phyciodes mylitta*) but latter is much brighter above. **WINGSPAN** 14 mm. **LARVA** Brown with fine white speckling and dark brown lateral band and bristly spines. **FOOD** Adult: Flowering plants (nectar). Larva: *Aster chilensis* and other asters (*Aster* spp.). **FOUND** Throughout the region, from coastal to mountain meadows.

Also known as *Phyciodes partensis* and *P. campestris*.

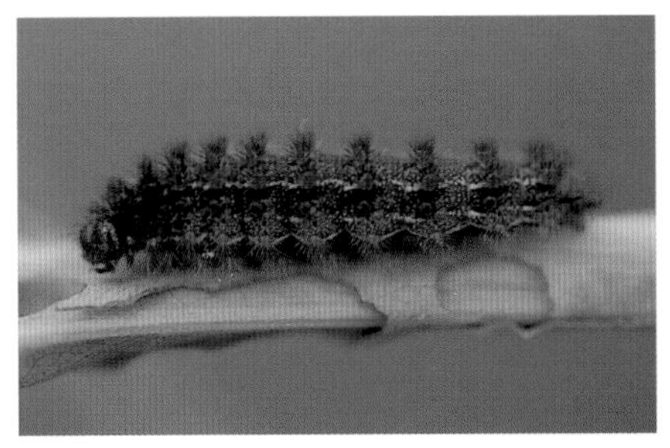
Phyciodes pulchellus, larva

Phyciodes pulchellus

Polygonia faunus

Polygonia faunus

Polygonia faunus, larva

Polygonia faunus

GREEN COMMA

ADULT Above, wings rich brownish orange with dark brown spots, and dark brown outer margin with submarginal row of pale yellow spots. Below, forewings brown and gray with green to bluish green submarginal band and spots; hindwings similar to forewings but each also with white comma hooked at both ends. **WINGSPAN** 50 mm. **LARVA** Banded orange and black with broad, white dorsal band and bristly spines. **FOOD** Adult: Flowering plants (nectar) and mammal feces. Larva: Willow (*Salix* spp.). **FOUND** Throughout the region.

Polygonia gracilis

Polygonia gracilis

HOARY COMMA

ADULT Above, forewings rich brownish orange suffused with golden yellow, and with dark brown spots and dark brown outer margin; hindwings similar to forewings but dark brown on outer margin reduced or missing. Below, forewings gray and brown with submarginal row of pale yellow spots; hindwings similar to forewings but each also with narrow, white comma tapered at both ends. **WINGSPAN** 50 mm. **LARVA** Black with pale orange reticulations, broad, white dorsal band, and bristly spines. **FOOD** Adult: Flowering plants (nectar) and mammal feces. Larva: Currant or gooseberry (*Ribes* spp.) (leaves). **FOUND** Throughout the region.

Also known as *Polygonia zephyrus*.

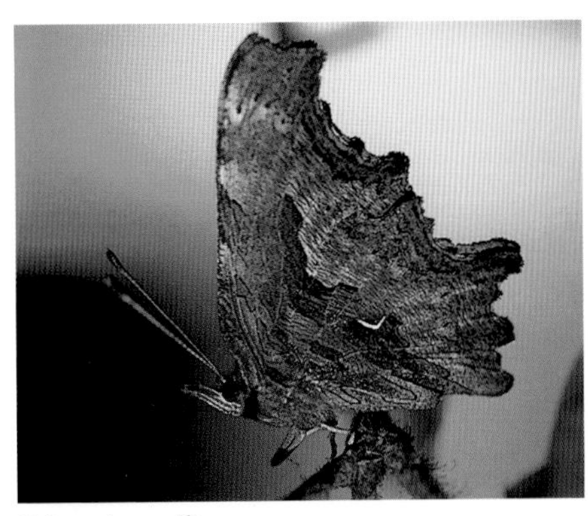

Polygonia gracilis

Polygonia gracilis, larva

Polygonia oreas

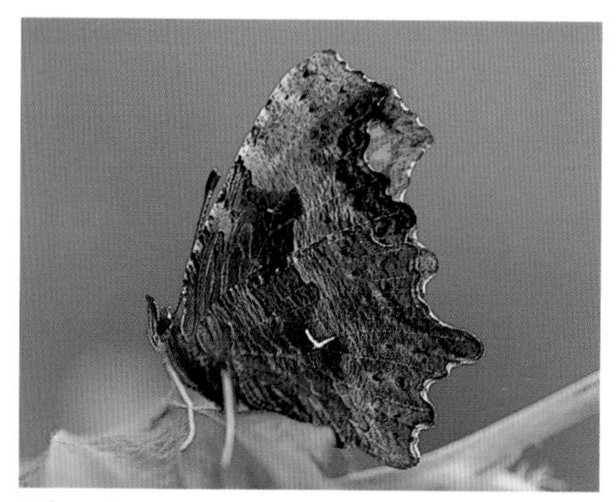

Polygonia oreas

Polygonia oreas, larva

Polygonia oreas

OREAS ANGLEWING

ADULT Above, wings orangish brown suffused with pale yellow, and with dark brown spots and outer margin with reduced, medium brown band. Below, forewings black, brown, and gray (with more dark brown than other anglewings); hindwings similar to forewings but each also with white comma tapered at both ends. **WINGSPAN** 50 mm. **LARVA** Black with pale orange reticulations, broad, white dorsal band, and bristly spines. **FOOD** Adult: Flowering plants (nectar) and mammal feces. Larva: Currant or gooseberry (*Ribes* spp.) (leaves). **FOUND** Throughout the region.

Polygonia satyrus

Polygonia satyrus

SATYR ANGLEWING

ADULT Above, forewings golden orange with dark brown spots and dark brown outer margin; hindwings similar to forewings but dark brown on outer margin reduced. Below, forewings mottled reddish brown; hindwings similar to forewings but each also with white comma clubbed at one end and hooked at other end. **WINGSPAN** 50 mm. **LARVA** Black with broad, white dorsal band (may be solid or broken into spots) and bristly spines. **FOOD** Adult: Flowering plants (nectar) and mammal feces. Larva: Stinging nettle (*Urtica* spp.). **FOUND** Very common throughout the region.

The more mature larva bites into the leaf stalk of a host plant, causing the leaf to droop. It then silks the leaf closed and uses it as a shelter.

Polygonia satyrus

Polygonia satyrus, larva

Speyeria cybele, male and female

Speyeria cybele, female

Speyeria cybele, larva

Speyeria cybele

GREAT SPANGLED FRITILLARY

ADULT Above, male, wings orange to bright orange with dark brown spots and bands and brown to dark brown basal area; female, wings dark purplish brown with wide, pale yellow submarginal band containing dark purplish brown spots. Below, male, forewings mostly creamy white with black spots and bars, and hindwings brown to dark brown with silvery white spots and wide, creamy white submarginal band; female, wings similar to male but less red. **WINGSPAN** 75 mm. **LARVA** Black with bristly, orange spines. **FOOD** Larva: Violet (*Viola* spp.). **FOUND** Throughout the region.

There is much variability within and between the species of this genus making it difficult to identify them to species. It takes years of patience and a butterfly net to properly identify them! One exception is *Speyeria cybele,* the female of which is so distinct that it cannot be confused with any other *Speyeria* spp. in the region. Another problem in identifying *Speyeria* spp. is that the larvae are very difficult to find: they often leave the host plant and hide when not feeding.

Speyeria hydaspe

HYDASPE FRITILLARY

ADULT Above, wings bright orange with brown base and black bars and spots. Below, forewings mostly orange with black spots and bars, and reddish brown and light purple at tip; hindwings reddish brown suffused with light purple, and with large, white spots (hindwing color separates this species from other *Speyeria* spp. when they are nectaring together). **WINGSPAN** 60 mm. **LARVA** Black with bristly spines. **FOOD** Larva: Violet (*Viola* spp.). **FOUND** Throughout the region.

This species is another exception to the rule that *Speyeria* spp. are difficult to identify (see *Speyeria cybele* for further information on this topic).

Speyeria hydaspe

Vanessa annabella

WEST COAST LADY

ADULT Above, forewings bright orange with black pattern and each with medium orange postmedial bar that touches leading margin; hindwings bright orange with black pattern and each with submarginal row of four black-ringed blue spots. **WINGSPAN** 50 mm. **LARVA** Variably patterned with black and yellow, and with bristly spines. **FOOD** Larva: Checker mal-

Vanessa annabella

Vanessa annabella, larva

Vanessa atalanta, larva

low (*Sidalcea malvaeflora*) and other checker mallows (*Sidalcea* spp.), stinging nettle (*Urtica* spp.), and many other herbaceous plants. **FOUND** Very common throughout the region.

This is the most common lady butterfly on the West Coast.

Vanessa atalanta

RED ADMIRAL

ADULT Above, forewings with dark brown base, orange medial band, and black with white spots at tip; hindwings dark brown with broad, orange band on outer margin. **WINGSPAN** 55 mm. **LARVA** Variably colored—most commonly black with fine white speckling, light yellow lateral spots, and bristly spines. **FOOD** Larva: Stinging nettle (*Urtica* spp.). **FOUND** Common throughout the U.S., wherever stinging nettles are found.

The larva in the later instars feeds by biting into the leaf stalk of a nettle, causing the leaf to droop. It then silks the leaf closed wherein it feeds.

Vanessa atalanta

Vanessa cardui

PAINTED LADY

ADULT Above, forewings orange with black pattern and each with postmedial white bar that touches leading margin; hindwings orange with black pattern and each with submarginal row of four small, black-ringed blue spots. Below, forewings mostly reddish pink with greenish brown, white, and black pattern; hindwings mottled with various shades of brown with pale white veins and each with row of four to five eyespots. **WINGSPAN** 55 mm. **LARVA** Variably colored—most commonly black with fine, white speckling, creamy white lateral line, and bristly spines. **FOOD** Larva: Thistle (*Cirsium* spp.), little mallow (*Malva parviflora*) and other mallows (*Malva* spp.), lupine (*Lupinus* spp.), and other plants. **FOUND** Throughout most of the U.S.

Despite the fact that this species is unable to overwinter in cold areas, it is the most common butterfly worldwide. It is usually

Vanessa cardui

Vanessa cardui, larva

Vanessa cardui

the second most common painted lady in the Northwest; it becomes the most common painted lady during the years when thousands of them pass through this region in the spring on their mass migration from Mexico northward. The most recent mass migration occurred in 2005. The adults who fly during these migrations are not as brightly colored as those who fly later in the season.

Vanessa virginiensis

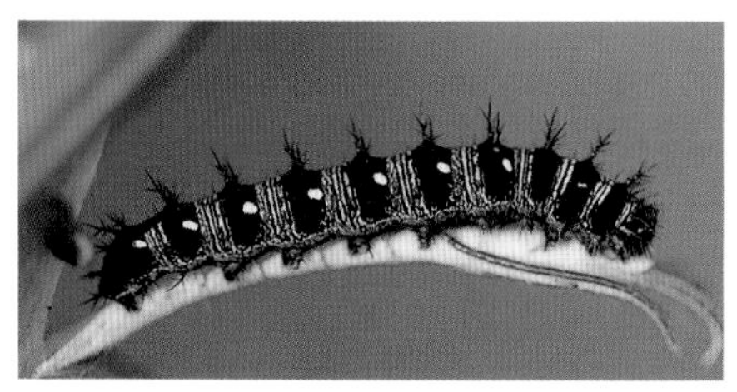

Vanessa virginiensis, larva

Vanessa virginiensis

AMERICAN LADY

ADULT Above, forewings reddish orange with black pattern and each with postmedial pale orange bar that touches leading margin; hindwings reddish orange with submarginal row of black and black-ringed blue spots. Below, forewings mostly dark pink with grayish brown, black, and white pattern; hindwings grayish brown with spidery network of creamy white lines and each with two large eyespots. **WINGSPAN** 54 mm. **LARVA** Black with yellow and black transverse bands and

Vanessa virginiensis

bristly spines. **FOOD** Larva: The pearly everlasting *Anaphalis margaritacea*, pussy-toes (*Antennaria* spp.), cudweed (*Gnaphalium* spp.). **FOUND** Throughout much of the U.S. Larva is very easy to find since it builds a conspicuous nest in the flower heads of the host plant.

This is the least common painted lady in the region.

SWALLOWTAIL & PARNASSIAN BUTTERFLIES

Family Papilionidae

This family is mostly represented by the swallowtail butterflies. They occur worldwide, with the greatest number of species in the tropics. Most of the swallowtails are large and brightly colored; in fact, this group includes the largest and some of the most beautifully colored butterflies in the world. Many swallowtail species are also easily recognized by the conspicuous "tails" on the posterior of the hindwings, for which this group gets its common name.

Another group that is included in this family is the parnassian butterflies, which look nothing like swallowtails but are taxonomically similar to them. These butterflies are medium in size and have wings that are usually white or gray with dark markings and red spots. They are mainly montane and boreal in distribution.

The larvae and pupae of swallowtails also differ from those of parnassians. Although some have fleshy tubercles, swallowtail larvae are usually smooth-skinned and hairless; parnassian larvae are pubescent. Swallowtails overwinter as pupae, which are cryptically colored and hang to objects by a silk girdle, whereas parnassians overwinter as eggs, emerge as larvae during the spring, and then pupate during the summer in loose cocoons in leaf litter on the ground.

One characteristic that all papilionid larvae have in common is the Y-shaped organ that protrudes from the thorax behind the head whenever the larva is threatened. This organ emits a malodorous substance, which acts as a defense mechanism against predators, particularly ants. Some species have prominent eyespots that combine with the snaketongue-like organ to resemble a snake's head, which also helps to ward off predators.

The adults of most species feed on flowers, and most larvae feed on foliage. In many cases, the adults and larvae are distasteful to vertebrate predators as a result of feeding on the host plants.

Battus philenor, mating pair

Battus philenor, larva

Battus philenor

PIPEVINE SWALLOWTAIL

ADULT Above, forewings black; hindwings iridescent blue. Below, hindwings with marginal half iridescent blue and with row of large orange spots. **WINGSPAN** 85 mm. **LARVA** Black with rows of red spinelike appendages; pair of long, spinelike appendages on back of head. **FOOD** Larva: The pipevine *Aristolochia californica*. **FOUND** Oregon and California, wherever pipevine naturally occurs.

The adult and larva are distasteful to birds. The larva's skin is so tough that the larva is often able to survive after having been sampled by a bird.

Papilio eurymedon

Papilio eurymedon

PALE TIGER SWALLOWTAIL

ADULT Above, wings white to creamy white with black bands and broad, black outer margin. Below, wings similar to above. **WINGSPAN** 95 mm. **LARVA** Very early stages, bird-dropping mimic; later stages, green with pair of yellow or orange eyespots with blue center. **FOOD** Larva: California-lilac (*Ceanothus* spp.), cascara (*Rhamnus purshiana*), California coffeeberry (*R. californica*), *Prunus* spp. **FOUND** Throughout the region.

Papilio eurymedon, larva

Papilio indra

Papilio indra, larva

Papilio indra

INDRA SWALLOWTAIL

ADULT Above, wings black with light yellow band and row of spots. Hindwing with short tail. **WINGSPAN** 85 mm. **LARVA** Black with light yellow transverse bands. **FOOD** Larva: *Cymopterus terebinthinus* and other members of carrot family (Apiaceae). **FOUND** Throughout the region, in rocky areas where host plant grows (larvae are often found under rocks when not feeding).

The larva of this species is one of the most beautiful swallowtail larvae.

Papilio multicaudatus

Papilio multicaudatus

TWO-TAILED TIGER SWALLOWTAIL

ADULT Above, forewings pale yellow with black bands (narrower than in *Papilio rutulus*) and black outer margin; hindwings similar to forewings. Hindwing with two tails. **WINGSPAN** 130 mm. **LARVA** Very early stages, bird-dropping mimic; later stages, green with pair of yellow eyespots with blue center; very similar to *Papilio eurymedon*. **FOOD** Larva: Plums and cherries (*Prunus* spp.), Oregon ash (*Fraxinus latifolia*) and other ashes (*Fraxinus* spp.), as well as other plants. **FOUND** Throughout the region.

This is the largest tiger swallowtail in the region.

Papilio rutulus

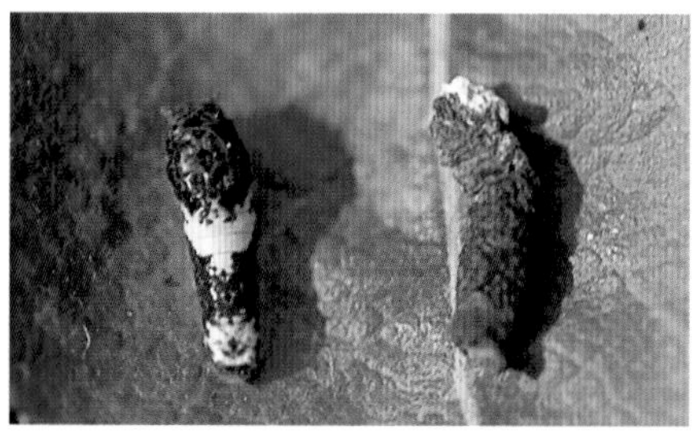

Papilio rutulus, larva and bird dropping

Papilio zelicaon, larva

Papilio rutulus

WESTERN TIGER SWALLOWTAIL

ADULT Above, forewings pale yellow with black bands (broader than in *Papilio multicaudatus*) and black outer margin; hindwings similar to forewings. **WINGSPAN** 98 mm. **LARVA** Very early stages, bird-dropping mimic; later stages, green with pair of yellow eyespots with blue center. **FOOD** Larva: Willow (*Salix* spp.), poplar (*Populus* spp.), sycamores (*Platanus* spp.), alder (*Alnus* spp.), and other plants. **FOUND** Throughout the region. Common in many habitats; most common tiger swallowtail in cool coastal areas.

Papilio zelicaon

ANISE SWALLOWTAIL

ADULT Above, forewings black with broad, pale yellow medial band; hindwings similar to forewings. **WINGSPAN** 75 mm. **LARVA** Very early stages, bird-dropping mimic; later

Papilio zelicaon

stages, green with broken transverse bands of black and yellow. **FOOD** Larva: Members of carrot family (Apiaceae) including *Lomatium* spp., yampah (*Perideridia* spp.), and fennel (*Foeniculum vulgare*). **FOUND** Throughout the region.

This species has greatly benefited from the spread of wild fennel.

Parnassius clodius

Parnassius clodius

CLODIUS PARNASSIAN

ADULT Above, forewings creamy to grayish white with grayish black bands, and black bars on leading margin; hindwings creamy to grayish white with grayish black bands and red spots. Below, wings same as above. Antennae black. **WINGSPAN** 75 mm. **LARVA** Velvety black with rows of yellow spots. **FOOD** Larva: Bleeding heart (*Dicentra formosa*) and other *Dicentra* spp. **FOUND** Throughout the region, from the coast to the mountains. Oc-

Parnassius clodius, larva

curs in same areas as *Parnassius smintheus* but usually flies at lower elevations. Larva is difficult to find when not feeding on host.

Parnassius smintheus

MOUNTAIN PARNASSIAN

ADULT Above, forewings creamy to grayish white with grayish black bands, black bars on leading margin, and red spots; hindwings similar to forewings but without black bars on leading margin. Below, wings same as above. Antennae banded black and white. **WINGSPAN** 75 mm. **LARVA** Black with yellow spots. **FOOD** Larva: *Sedum* spp. **FOUND** Throughout the region. Occurs in same areas as *Parnassius clodius* but usually flies at higher elevations.

Also known as *Parnassius phoebus*. Pete thinks this species is one of the most beautiful butterflies on the West Coast.

Parnassius smintheus

SULPHUR, WHITE & ORANGE-TIP BUTTERFLIES

Family Pieridae

This family includes many familiar butterflies. Many species are very common and abundant. Pierids occur in a wide range of habitats, from deserts and coasts to timberline. The adults of most species are medium in size and some shade of white or yellow, usually with dark markings; some species also have orange markings. A characteristic unique to the Pieridae is that they may occur in different color forms, depending on the season. The adults are strong flyers, with many having a rapid, straight-lined flight. They nectar on flowers and can also be found puddling (feeding on moist or wet soils). The larvae are somewhat elongate with smooth, usually green skin covered with short hairs. In most species, they feed on plants of either the mustard or legume families. The mustards contain mustard oils that are distasteful to many vertebrate predators. Consequently, predators learn to avoid the larvae that feed on those plants. The pupae are elongate and attached to objects by a silk girdle.

Anthocharis lanceolata

GRAY MARBLE

ADULT Above, forewings white, each with black spot. Below, forewings white with mottled gray tip; hindwings mottled with gray. **WINGSPAN** 44 mm. **LARVA** Green with white speckling and white lateral line. **FOOD** Larva: *Arabis glabra* and other rock cresses (seed pods). **FOUND** Oregon and California.

Also known as *Paramidea lanceolata*.

Anthocharis lanceolata

Anthocharis sara, pupa

Anthocharis sara

SARA'S ORANGE-TIP

ADULT Above, forewings white with distinct orange tip bordered with black. Below, hindwings white with yellow veins and grayish green mottling. **WINGSPAN** 40 mm. **LARVA** Green with black speckling and creamy white lateral line. **PUPA** Shaped like a thorn; green or brown. **FOOD** Larva: *Arabis glabra* and other rock cresses (*Arabis* spp.), as well as other cruciferous plants (usually flowers and seed pods). **FOUND** Throughout the region.

The start of the flight season for this species is early spring.

Anthocharis sara

Colias eurytheme

ORANGE SULPHUR

ADULT Above, wings orange or yellowish orange with black outer margin. Below, wings pale orange, yellow, or greenish yellow and usually edged in pink. Very similar to the western sulphur (*Colias occidentalis*) (not described in this field guide) except that the latter is yellow with no orange. **WINGSPAN** 51 mm. **LARVA** Green with white, yellow, and red lateral line. **FOOD** Larva: *Lupinus rivularis, L. polyphyllus,* and other lupines (*Lupinus* spp.); alfalfa (*Medicago sativa*). **FOUND** Throughout the U.S. Found in pristine mountain meadows or by the thousands in alfalfa fields.

This native insect has adapted well to the introduction of European legumes.

Colias eurytheme

Colias eurytheme

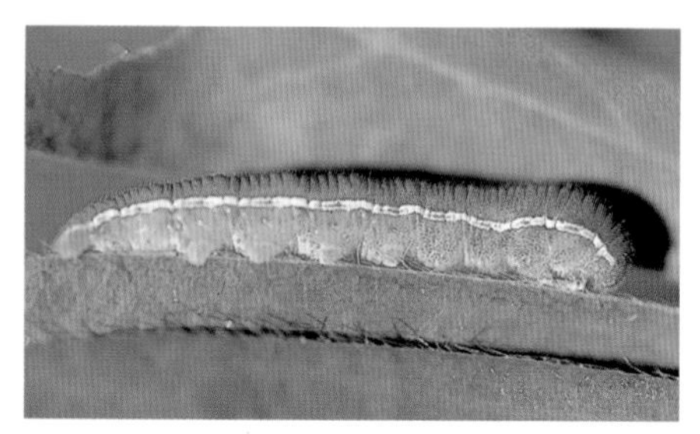

Colias eurytheme, larva

Euchloe ausonides

LARGE MARBLE

ADULT Above, forewings white, each with small, black medial patch and black barring at tip; hindwings white. Below, forewings white with yellow and grayish green mottling at tip; hindwings white with yellow along veins and grayish green mottling. **WINGSPAN** 44 mm. **LARVA** Grayish blue above, green below, with yellow longitudinal lines and small, raised, black spots. **FOOD** Larva: *Arabis glabra* and other rock cresses (*Arabis* spp.), as well as other mustards. **FOUND** From Alaska to California.

This is a difficult butterfly to observe in the wild because it flies erratically, low to the ground, and only for a short time.

Euchloe ausonides

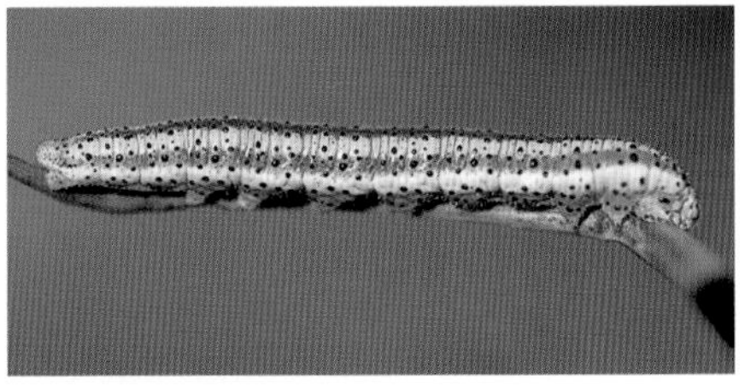

Euchloe ausonides, larva

Neophasia menapia

PINE WHITE

ADULT Below, male, forewings white, each with black bar on leading margin and black veins and band (or patch) at tip, and hindwings white with black veins; female, forewings white, each with black bar on leading margin and black veins and band (or patch) at tip, and hindwings same as male but veins more heavily black and margins edged in red. **WINGSPAN** 50 mm. **FOOD** Larva: Douglas-fir (*Pseudotsuga menziesii*), pine (*Pinus* spp.). **FOUND** Common throughout the region. The habitat in which Pete has found the highest number of adults has been old growth Douglas-fir forests.

The adult flies later in the flight season than do other native whites; when it descends from tree tops, it looks like a piece of gauze since it floats more than flies. The adult is very delicate and should be handled very carefully.

Neophasia menapia, male

Neophasia menapia, female

Pieris napi

Pontia beckerii

Pieris napi

VEINED WHITE

ADULT Above, wings white. Below, forewings creamy white with veins edged with grayish black; hindwings similar to forewings but veins usually more heavily edged with grayish black. **WINGSPAN** 18 mm. **FOOD** Larva: Mustard family (Brassicaceae). **FOUND** Throughout most of the northern U.S.

Also known as *Pieris marginalis.*

Pontia beckerii

BECKER'S WHITE

ADULT Below, forewings white, each with black rectangle in middle and yellow veins bordered with grayish green at tip; hindwings white with yellow veins bordered with grayish green. **WINGSPAN** 45 mm. **FOOD** Larva: Mustard family (Brassicaceae). **FOUND** Throughout the region.

Also known as *Pieris beckerii.*

BAGWORM MOTHS Family Psychidae

This is a small family of moths in which the small, inconspicuous adult males have well-developed wings, which are black or transparent in color, and the adult females of most species are wingless, legless, and larviform. The common name for the family is derived from the characteristic case, or "bag," that is constructed and carried around by the larva while feeding on plants. This spindle-shaped, silken case is covered with bits of leaves, twigs, or other debris. Both ends of the case are open, one end for walking and feeding and the other for fecal discharge. The larva enlarges its case as it grows. When full grown, the larva attaches its case to a substrate with silk, closes it, and pupates inside. The males emerge from their pupal case after maturing, while the females typically remain in their case and attract males with which to mate by emitting pheromones from their abdomen. Upon emerging, a male will fly to a female's case and thrust his abdomen through the posterior end of the case. After fertilization, the female lays her eggs in the case. Both adults, who do not feed, die shortly after the eggs are laid. After hatching, the larvae crawl away from the case to feed and to form their own cases.

Hyaloscotes fumosa

ADULT Male, wings brownish gray; body gray and hairy. **WINGSPAN** Male, 18 mm. **FOOD** Larva: *Eriogonum compositum* and other wild buckwheats (*Eriogonum* spp.), and possibly other plants. **FOUND** Range unknown, but this species probably occurs throughout the region, at higher elevations.

Hyaloscotes fumosa, bag

Hyaloscotes fumosa, newly emerged male and bag with pupal skin

SNOUT MOTHS Family Pyralidae

This is the second-largest family of lepidopterans, and many species are very common. These moths greatly vary in appearance and habits. The adults of most species are small with forewings that are elongate or triangular and hindwings that are usually broad. They hold their wings out to the side, fold them flat, or roll them up, making their bodies look like sticks. The common name of the family is derived from the elongated "snout" of the adult. The habits of the larvae greatly vary: some feed on leaves, either skeletonizing or rolling the leaves or forming webs on them; others bore into roots or stems, scavenge, prey on other insects, or feed on moss. Some species are pests of crops, stored food products, and beehives.

Pyrausta californicalis

Pyrausta californicalis

CALIFORNIA PYRALID

ADULT Above, forewings reddish brown, each with pale yellow band and spot. **WINGSPAN** 12 mm. **FOOD** Larva: Unknown. **FOUND** Oregon and California.

METALMARK BUTTERFLIES Family Riodinidae

This is a large, diverse family of butterflies that occurs worldwide but is predominantly tropical. There is only one species in the Pacific Northwest. The common name of the family originates from the metallic markings on the wings of many species. Compared to the tropical members, which are characterized by brilliant colors, most species found in North America are relatively subtle in color. The adults frequently rest and feed with their wings opened nearly flat against the substrate; many rest head downward on the undersides of leaves. All species nectar on flowers. The larvae are short, sluglike, and somewhat pubescent. The pupae are stout and often covered with short hair and attached by a silk girdle to leaf litter or to the stem of the host plant.

Apodemia mormo

MORMON METALMARK

ADULT Above, wings variably colored and patterned, usually a combination of orange and dark brown with white spots and dark brown margins. Below, forewings with orange base, brown to gray tip, and white spots; hindwings brown to gray with white spots. **WINGSPAN** 32 mm. **LARVA** Purple with clumps of short hairs arising from black bases completely or partially ringed with yellow. **FOOD** Larva: Sulfur flower (*Eriogonum umbellatum*) and other wild buckwheats (*Eriogonum* spp.). **FOUND** Throughout the region.

Apodemia mormo

Apodemia mormo, larva

GIANT SILKWORM MOTHS Family Saturniidae

The members of this primarily tropical family are medium- to large-sized moths of which many are richly colored and strikingly patterned, some having large, transparent eyespots on the wings; this family includes some of the largest and most beautiful lepidopterans in the world. Nine species occur in the Pacific Northwest. These moths have stout, short bodies covered with hairlike scales that make them appear furry; the female is larger than the male. The antennae are large and feathery and conspicuously larger in the male than in the female. The males use their antennae to sweep the air to detect pheromones released by sexually receptive females to whom they are attracted. The mouthparts are reduced, and the adults do not feed. They are active during the day or at night, depending on the species. Nocturnal males are attracted to lights at night.

The larvae are large and usually have conspicuous bumps or spines. Some have urticating hairs that protect them from predators and from handling by curious humans. The larvae feed mostly on the foliage of trees and shrubs. They are usually easier to find than the adults. A partially eaten leaf or piles of feces on leaves or on the ground below the host plant are indications that a larva is present. A host plant with the tips of its branches stripped of leaves indicates the presence of a group of feeding larvae. Many spin a strong, dense silken cocoon (hence the *silkworm* part of the common name for the family), which may be attached to a twig or hidden in fallen leaves; others pupate in a cell in the soil. Most overwinter as pupae. This family is not closely related to the family that includes the silkworm found in Asia.

Antheraea polyphemus, larva

Antheraea polyphemus, newly emerged adult

Antheraea polyphemus

POLYPHEMUS MOTH

ADULT Above, forewings pale brown to reddish brown, each with eyespot containing translucent center; hindwings similar to forewings except eyespot larger and surrounded by black and blue. Below, wings banded brownish gray and brown. **WINGSPAN** 120 mm. **LARVA** Green with yellow lateral lines and small, red bumps. **FOOD** Larva: Oak (*Quercus* spp.), maple (*Acer* spp.), and other woody broadleaf trees and shrubs. **FOUND** Throughout North America; every state in the U.S., except Nevada and Arizona.

This nocturnal moth is the most widely distributed giant silkworm moth in North America and one of the largest moths in the region, having a wingspan that can reach 140 mm. The larva feeds singly and, until it has

Antheraea polyphemus

grown to a large size, can be very difficult to find on the host plant (look for fecal pellets or chewed leaves). In later instars, it can hide its feeding activity by "stemming" the damaged leaf, causing it to fall to the ground. In the winter Pete has found cocoons at the base of their host plants.

Antheraea polyphemus

Coloradia pandora

PANDORA MOTH

ADULT Above, forewings brown suffused with gray and each with black spot and pair of black jagged lines in middle of wing; hindwings much lighter in color and tinged with pink. **WINGSPAN** 80 mm. **LARVA** Dark greenish brown with white spots, white longitudinal dashed bands, and light yellow transverse bands at posterior of each segment; covered with short black spines. **FOOD** Larva: Jeffrey pine (*Pinus jeffreyi*), Pacific ponderosa pine (*P. ponderosa*), lodgepole pine (*P. contorta*). **FOUND** Oregon and California.

This species normally has a two-year life cycle. The first-year larvae overwinter on the needles of the host trees. During the follow-

Coloradia pandora, larva

ing spring, with the return of warmer weather, the larvae eat until they mature. They then pupate in the soil during the summer. The adults emerge from the pupae usually the next spring or summer, although some adults may take longer to emerge. This is the only giant silkworm moth in the region that may cause major economic loss to commercial forests.

Hemileuca eglanterina, larva

Hemileuca eglanterina

SHEEP MOTH

ADULT Above, forewings yellowish orange suffused with pink and each with black medial spot, bands, and marginal bars; hindwings similar to forewings but without pink. **WINGSPAN** 65 mm. **LARVA** Brown or black with pale white lateral lines and red transverse lines and covered with rows of bristly, black spines, the two dorsal rows of which are with distinct yellow to orange bases. **FOOD** Larva: Blue blossom (*Ceanothus thyrsiflorus*), *C. velutinus*, mountain whitethorn (*C. cordulatus*), deer brush (*C. integerrimus*), and other California-lilacs (*Ceanothus* spp.); Sierra gooseberry (*Ribes roezlii*) and other currants or gooseberries (*Ribes* spp.); willow (*Salix* spp.); the mountain-mahogany *Cercocarpus betuloides*; the antelope brush *Purshia tridentata*; and other plants. **FOUND** Throughout the region.

This is probably the most common giant silkworm moth in the region. Because they

Hemileuca eglanterina

are so colorful and fly during the day, the adults are easily mistaken for butterflies. The larvae have urticating hairs that may irritate human skin (although by carefully handling the larvae, one may avoid this problem). They are the easiest to find of all the local species because their late instar stages are darkly colored and they feed in groups, which make them stand out against the branches and leaves of a host plant.

Hyalophora euryalus

Hyalophora euryalus

CEANOTHUS SILK MOTH

ADULT Above, forewings rich red to reddish brown, each with white, comma-shaped medial spot; hindwings similar to forewings except medial spot much larger. Below, wings similar to above but grayer. **WINGSPAN** 130 mm. **LARVA** Late instars usually green to pale green with rows of white to light yellow spines, with first two or three pairs of dorsal spines banded with black and yellow. **FOOD** Larva: Blue blossom (*Ceanothus thyrsiflorus*),

Hyalophora euryalus, larva

Hyalophora euryalus

Saturnia albofasciata, female

Saturnia albofasciata, larva

Saturnia mendocino

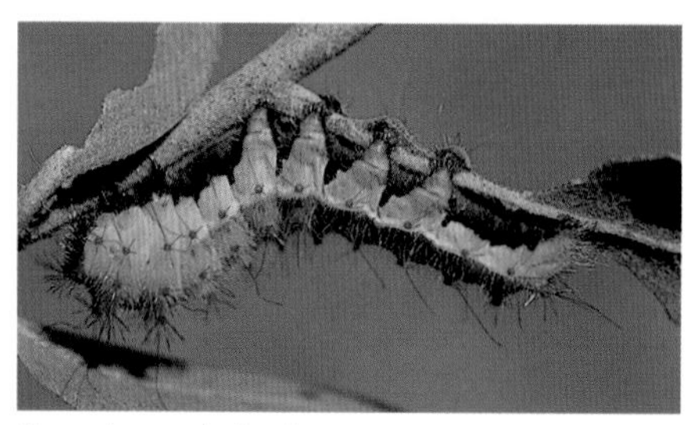

Saturnia mendocino, larva

C. velutinus, mountain whitethorn (*C. cordulatus*), deer brush (*C. integerrimus*), and other California-lilacs (*Ceanothus* spp.); Douglas-fir (*Pseudotsuga menziesii*); manzanita (*Arctostaphylos* spp.); and other plants. **FOUND** Common throughout the region.

Because of the distinctive comma on the hindwing, a new common name has been coined for this species—the *Nike moth*. The adults are nocturnal.

Saturnia albofasciata

ADULT Above, male, forewings brown, each with white medial band and large eyespot, and hindwings brownish orange, each with less distinct, white medial band and large eyespot; female, wings grayish brown, each with white medial band and large eyespot. **WINGSPAN** 27 mm. **LARVA** Green above, dark brown below, covered with white hairs, and with black spines arising from bright pink bumps. **FOOD** Larva: *Ceanothus cuneatus* and other California-lilacs (*Ceanothus* spp.), as well as other plants. **FOUND** California.

In the adults, the female is nocturnal, and the male is diurnal.

Saturnia mendocino

MENDOCINO SILK MOTH

ADULT Above, forewings brown to reddish brown with leading margin reddish orange and each with medial eyespot partially surrounded with white; hindwings orange, each with black postmedial band and medial eyespot. **WINGSPAN** 35 mm. **LARVA** Early instar stages variable in color. Last instar, green above, brown below, covered with white hairs and few, long, black hairs; with black spines arising from red bumps and pale yellow lateral line. **FOOD** Larva: Manzanita (*Arctostaphylos* spp.), Pacific madrone (*Arbutus menziesii*). **FOUND** Oregon and California.

The adult flies during the day.

CLEAR-WINGED MOTHS Family Sesiidae

These small- to medium-sized, slender moths are called clear-winged moths or wasp moths because many species in this family bear a strong resemblance to wasps—they have mostly transparent wings, long and narrow forewings, and brightly colored bodies, some having black bodies banded with yellow or red. Some species even mimic the threatening posture of wasps. Their wasplike appearance and behavior have probably evolved in response to vertebrate predators that learn to avoid stinging insects. The adults of most species are day-flyers and frequent flowers. The larvae bore into a variety of woody and herbaceous plants. Some are pests of fruit and ornamental trees.

Albuna pyramidalis

ADULT Above, forewings iridescent red with black veins. Abdomen banded with yellow and black. **WINGSPAN** 40 mm. **FOOD** Adult: Unknown (Pete has found them on mountain ash [*Sorbus* spp.]). Larva: Unknown. **FOUND** Probably throughout the region, in the mountains.

Albuna pyramidalis

Pennisetia marginatum

RASPBERRY CROWN BORER

ADULT Yellow jacket mimic—wings transparent with dark brown margins; abdomen black with yellow bands. **WINGSPAN** 16 mm. **FOOD** Larva: *Rubus* spp. (cane borer). **FOUND** Throughout the region.

Pennisetia marginatum

Synanthedon bibionipennis, female

Synanthedon polygoni

Synanthedon sequoiae, larva with pitch mass

Synanthedon bibionipennis

STRAWBERRY CLEARWING

ADULT Yellow jacket mimic—wings transparent, highlighted with gold, and with dark brown veins and margins; abdomen black with three (male) or four (female) yellow bands. **WINGSPAN** 22 mm. **FOOD** Larva: Beach strawberry (*Fragaria chiloensis*) and other strawberries (*Fragaria* spp.), *Rubus* spp. **FOUND** Throughout the U.S.

Synanthedon polygoni

ADULT Wings iridescent dark blue with orangish red on leading margin; abdomen iridescent dark blue with orangish red bands and tuft of orangish red hairs at tip. **WINGSPAN** 15 mm. **FOOD** Adult: Flowering plants. Larva: Wild buckwheat (*Eriogonum* spp.) (roots). **FOUND** Throughout the region. Look for adult in early spring nectaring at or near host plant.

Synanthedon sequoiae

SEQUOIA PITCH MOTH

ADULT Yellow jacket mimic—wings transparent with dark brown margins; abdomen black with yellow bands. Similar to *Pennisetia marginatum*. **WINGSPAN** 25 mm. **LARVA** Mostly hairless; head pale brown; abdomen white. **LARVA LENGTH** 24 mm. **FOOD** Larva: Lodgepole pine (*Pinus contorta*), Monterey pine (*P. radiata*), and other pines (*Pinus* spp.); Douglas-fir (*Pseudotsuga menziesii*). **FOUND** Throughout the region.

Also known as *Vespamima sequoiae*. The larva feeds in the cambium layer of the host tree, causing large amounts of pitch to accumulate on the bark (see photograph).

HAWK MOTHS Family Sphingidae

The members of this mostly tropical family are medium- to large-sized moths that have stout, spindle-shaped bodies. The elongate forewings are often cryptically colored, while the smaller hindwings are often boldly or colorfully patterned. These moths are strong flyers and have a very rapid wing beat; in fact, they are among the fastest flyers in the Lepidoptera. Some of the species resemble hummingbirds or large bees in flight. Most species also feed like hummingbirds, hovering in front of a flower and extending their tongue into it. Many species have a very long tongue, sometimes as long as or longer than the body, which enables them to nectar in deep-throated flowers. Depending on the species, the adults are either diurnal, nocturnal, or active at dusk or dawn.

The larvae are large, stout, and usually smooth-skinned. Many of the larvae are commonly called *hornworms* because of the prominent dorsal horn at the tip of the abdomen. Members of this family are also commonly called *sphinx moths*, which probably arose from the defensive posture of the larvae of some species—they raise the front part of the body while keeping the head tucked down, resembling the form of an Egyptian sphinx. The larvae feed both day and night on a wide variety of woody and herbaceous plants. Most pupate in a cell in the ground. Although the larvae of some species are considered pests, sphingids are important plant pollinators.

Hemaris diffinis

BUMBLE BEE MOTH

ADULT Above, forewings transparent with reddish black at base, black veins, outer margin bordered with black, and dark red at tip. Thorax covered with yellow hairs. Abdomen black with broad, yellow band. **WINGSPAN** 37 mm. **LARVA** Counter-shaded—green above, dark brown below; with brown anal horn. **FOOD** Larva: *Lonicera hispidula* and other honeysuckles (*Lonicera* spp.), as well as other related plants. **FOUND** Throughout the region.

The adult hovers like a hummingbird while nectaring. When an adult first emerges, its wings are solid black; after its initial flight, most of the black scales covering the wings are lost.

Hemaris diffinis

Hemaris diffinis, larva

Hyles lineata

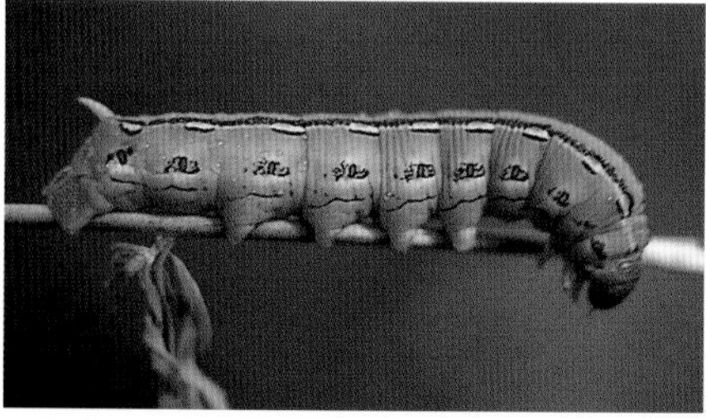

Hyles lineata, larva

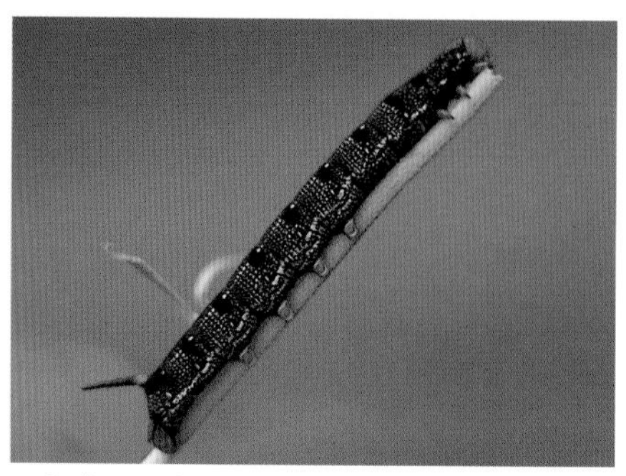

Hyles lineata, larva

Hyles lineata

WHITE-LINED SPHINX

ADULT Above, forewings dark brown with light brown band extending from base to tip and crosshatched with white lines; hindwings pink with black basal and marginal bands. **WINGSPAN** 85 mm. **LARVA** Two color variations—green with thin, black lines, and often with subdorsal row of pale yellow spots, or black with yellow lines; with anal horn. **LARVA LENGTH** 90 mm. **FOOD** Larva: Fireweed (*Epilobium angustifolium*), *E. ciliatum*, and other fireweeds (*Epilobium* spp.); and many other plants. **FOUND** Throughout the U.S., commonly in gardens and urban areas. Adult flies both at night and during the day, but is most often seen at dusk, nectaring like a hummingbird, at long-necked flowers.

Proserpinus clarkiae

CLARK'S DAY SPHINX

ADULT Above, forewings pale green and gray with broad, dark greenish brown medial band; hindwings orange with outer margin bordered with black. **WINGSPAN** 38 mm. **LARVA** Pale pink (sometimes faint) with wide, black dorsal band, large, black lateral spots, and flat anal button. **FOOD** Larva: Fireweed (*Epilobium angustifolium*), *E. brachycarpum*, and other fireweeds (*Epilobium* spp.); *Clarkia* spp. **FOUND** Throughout the region.

The adult flies in early spring.

Proserpinus clarkiae

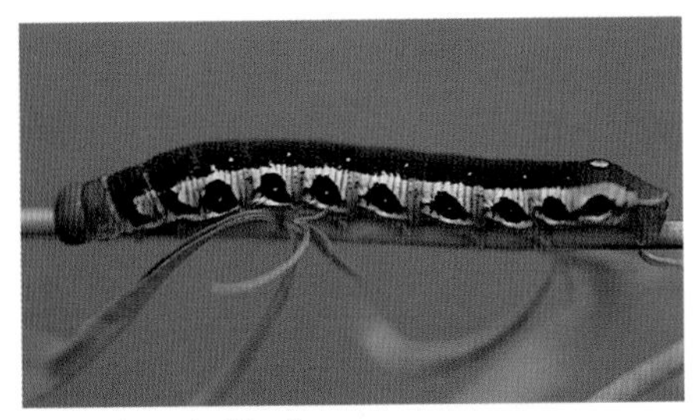

Proserpinus clarkiae, larva

Proserpinus flavofasciata

YELLOW-BANDED DAY MOTH

ADULT Bumble bee mimic—above, forewings black with white medial band; hindwings yellow with black basal and marginal bands; head and thorax yellow; abdomen black with small yellow patches near tip. **WINGSPAN** 38 mm. **LARVA** Later stages, dark brown with small, black spots and flat anal button consisting of black center ringed with pink and black. **FOOD** Larva: Fireweed (*Epilobium angustifolium*) and probably other fireweeds (*Epilobium* spp.). **FOUND** Throughout the region. Adult flies in early spring. Later stage larva often hides at base of host plant when not feeding.

Proserpinus flavofasciata

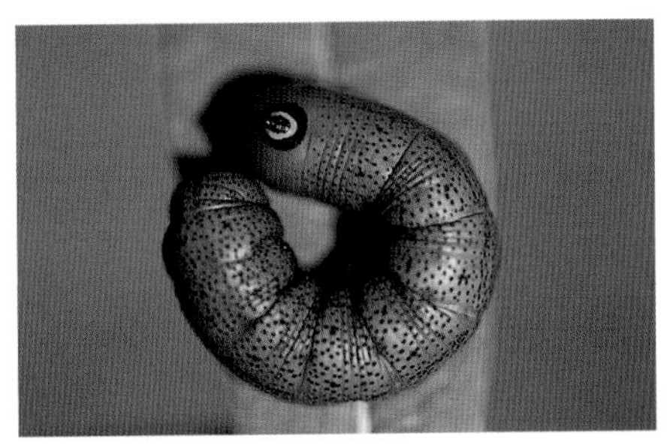

Proserpinus flavofasciata, larva

Smerinthus cerisyi

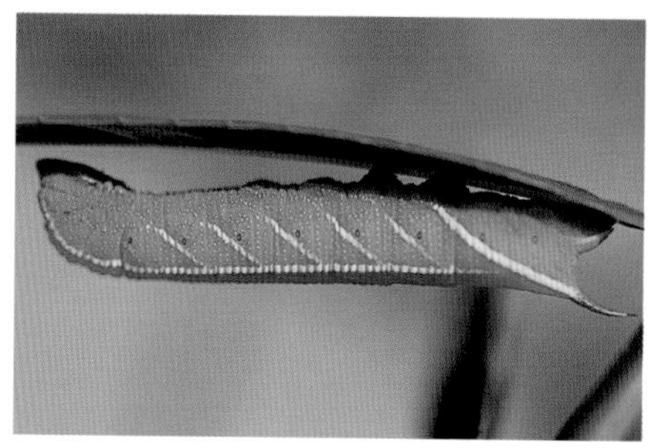

Smerinthus cerisyi, larva

Smerinthus cerisyi

EYED SPHINX

ADULT Above, forewings brownish gray with dark brown lines and bands and hooked tip; hindwings pinkish red, each with eyespot. **WINGSPAN** 75 mm. **LARVA** Green with light yellow subdorsal line, oblique, pale yellow lateral lines, and pink, blue, and/or yellow anal horn. **FOOD** Larva: Poplar (*Populus* spp.), willow (*Salix* spp). **FOUND** Throughout the region. Adult male is often attracted to lights at night in the spring and early summer.

This is one of the most common sphinx moths in the region. When at rest, the adult holds its wings closed, exposing only its forewings and thus camouflaging itself; when it is threatened, it raises its forewings to expose the eyespot on the hindwings.

Sphinx chersis

CHERSIS SPHINX

ADULT Above, forewings gray with black streaks; thorax gray with two black lines. **WINGSPAN** 90 mm. **FOOD** Larva: Oregon ash (*Fraxinus latifolia*), *Prunus* spp., and other trees and shrubs. **FOUND** Throughout the region.

Sphinx chersis

Sphinx drupiferarum

WILD CHERRY SPHINX

ADULT Above, forewings dark brown with black streaks and grayish white leading and outer margins. **WINGSPAN** 100 mm. **LARVA** Green with seven oblique, purple-and-white lateral lines and purple anal horn. **FOOD** Larva: Bitter cherry (*Prunus emarginata*) and other *Prunus* spp. **FOUND** Throughout most of the U.S.

Sphinx drupiferarum

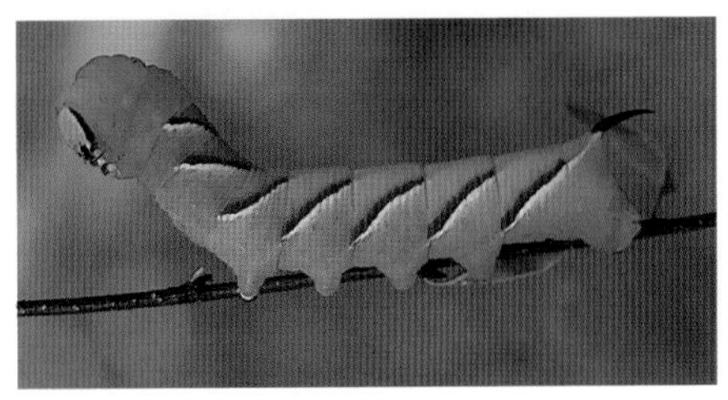

Sphinx drupiferarum, larva

Sphinx vashti

SNOWBERRY SPHINX

ADULT Above, forewings gray with long, black postmedial line and black line from tip. **WINGSPAN** 80 mm. **LARVA** Pale green with seven oblique, black-edged white lines on each side and three yellow bands with raised white bumps immediately posterior to head; large black and pale green posterior horn. **FOOD** Larva: Snowberry (*Symphoricarpos albus*) and other snowberries (*Symphoricarpos* spp.). **FOUND** Throughout the region. Larva blends in with foliage and moves very little, so is very difficult to see on host plant; look for last instar where feeding on leaf is apparent.

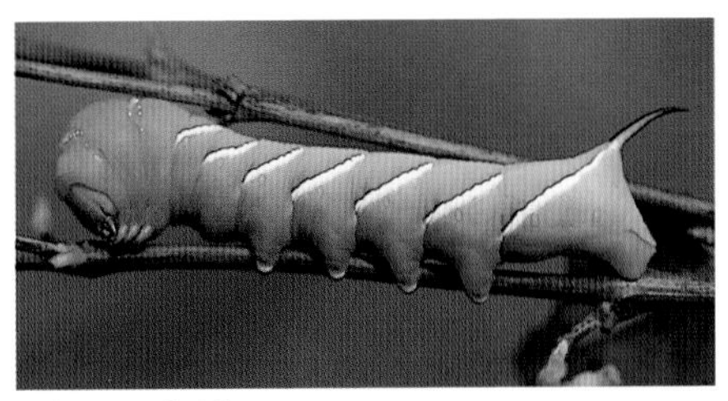

Sphinx vashti, larva

THYATIRID MOTHS Family Thyatiridae

The North American species of this small family of medium-sized moths often have distinctly patterned forewings. They resemble noctuid moths (Noctuidae) but have different wing venation. The larvae live and feed in loosely rolled leaves on trees and shrubs.

Habrosyne scripta

Habrosyne scripta

ADULT Above, forewings gray-brown, each with black, brown, and whitish gray zigzag pattern through middle. **WINGSPAN** 35 mm. **FOOD** Larva: Perennial broadleaf plants. **FOUND** Throughout the region. Adult is sometimes attracted to lights at night.

TORTRICID MOTHS Family Tortricidae

These small moths are usually brown or gray with spots or mottling. In most species the forewings are wide-angled at the tip. When at rest, the wings are usually held flat over the abdomen, which gives the moth a bell-shaped outline. The adults of most species are nocturnal; a few are day-flyers. The adults do not feed. The larvae vary in habit: many species roll or tie leaves, and others bore into stems, roots, or fruits. Most feed on perennial plants. Many species pupate in the leaf nests they created as feeding larvae or spin cocoons in leaf debris or under bark. Some species are orchard or forest pests.

Choristoneura rosaceana

OBLIQUE-BANDED LEAFROLLER

ADULT Above, forewings light reddish brown with many fine, dark reddish brown lines and wide, dark reddish brown medial band. **WINGSPAN** 20 mm. **LARVA** Pale green; head black. **EGG** Pale green; laid in flat clusters, overlapping each other like fish scales. **FOOD** Larva: Broadleaf trees and shrubs. **FOUND** Throughout the region.

In the spring the larva often feeds inside two leaves silked together or a leaf silked to the side of an apple or pear fruit (it causes superficial damage to the fruit).

Choristoneura rosaceana

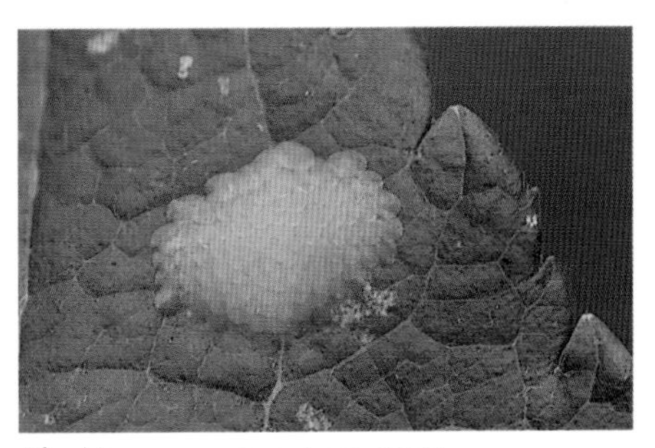

Choristoneura rosaceana, egg mass

ANTLIONS, SNAKEFLIES, AND ALLIES
Order Neuroptera

Neuroptera (Greek, *neuro*, sinew; *ptera*, wings) is a small order, with approximately 4670 known species worldwide. Approximately 350 species have been recorded in North America. Although the members of this order occur in almost all regions of the world, they are not very well known.

Adult neuropterans are medium to large in size and have two pairs of membranous wings, which, in most species, are elongate-oval in shape and similar in size, usually longer than the abdomen. The wings characteristically have many-branched, longitudinal veins connected by a great many cross veins. The net formed by these crisscrossing veins is thought to account for the name of the order, which is based on the use of the word *nervation*, meaning *strengthening of sinews*. Nevertheless, the adults of most species are not strong flyers. When at rest, the wings are usually held rooflike over the abdomen. The body of the adult is soft, elongate and cylindrical in shape, and usually brown. The legs are well developed, and the antennae are usually long and many-segmented. The mouthparts are of the chewing type, except in some species, in which the adults do not feed.

Neuropterans undergo complete metamorphosis. The bodies of the larvae vary in shape, from short and oval to elongate, but are generally flattened in appearance. In all species, the larvae have chewing mouthparts, often with large, sickle-shaped jaws.

Most species of neuropterans are terrestrial. The adults are usually found near their larval habitat. The adults of some species are predaceous, while those of other species feed on nectar and pollen. The larvae of most species are predaceous, and some are parasitic. The larval members of one family feed exclusively on freshwater sponges.

Since most species are predaceous, at least in the larval stage, neuropterans are important in regulating invertebrate populations. In fact, some species are used for the biological control of small insect pests of crops.

ANTLIONS Family Myrmeleontidae

This is the largest family of neuropterans. Most of the species occur in the more arid regions of the world. The adults look very similar to damselflies with their long, slender abdomen and long, narrow, many-veined wings. They differ from the damselflies in having soft bodies, different wing venation, and longer, clubbed antennae. The wings are transparent or mottled with black or brown. The adults are weak flyers. Most species are nocturnal or fly at dusk and often are attracted to lights. The adults feed on nectar or pollen or do not eat at all. The cryptically colored larvae are stout-bodied with huge sicklelike jaws. All are predaceous. The family gets its common name from the fact that some species construct a funnel-shaped pit trap in sandy or fine soil and wait at the bottom to catch ants and other small insects that fall down into the pit. Other species lie buried in the sand or hide among debris waiting for prey. Antlions pupate in parchmentlike cocoons buried in the sand.

Brachynemurus ferox

ADULT Brown and light gray. **BODY LENGTH** 40 mm. **FOOD** Larva: Ants and other invertebrates. **FOUND** Throughout the region. Larva, in sandy or fine soil at bottom of funnel-shaped pit.

Since the adult is a very weak flyer, it can be easily caught by hand.

Brachynemurus ferox, larva

Brachynemurus ferox

SQUARE-HEADED SNAKEFLIES Family Inocelliidae

This family of very distinctive (some might even say peculiar) insects is found, within the United States, only in the western states. One obvious characteristic that separates these snakeflies from other snakeflies is their square head. As in other snakefly families, adult square-headed snakeflies have an elongate, necklike prothorax and long, transparent, intricately veined wings. When at rest, they hold their wings rooflike over the body and elevate their head. Snakefly adults are active predators but capable of catching only small and weak prey. The larvae are usually found under bark, feeding mainly on small soft-bodied insects. Snakefly families are sometimes put in a separate order, Raphidioptera.

Negha inflata

Negha inflata

ADULT Wings with black veins; abdomen black with yellow bands; antennae very long and thin. **BODY LENGTH** 25 mm. **FOOD** Adult and larva: Invertebrates. **FOUND** Throughout the region.

The adult moves very slowly and can be easily caught by hand.

DRAGONFLIES AND DAMSELFLIES Order Odonata

Odonata (Greek, *odon*, tooth) is a small order of familiar aquatic insects, with only 4870 known species worldwide, and 430 of those species recorded in the United States and Canada. This order is of ancient lineage: fossils of giant dragonflies that date back approximately 250 million years have wingspans of over two feet, making them some of the largest known insects. The distribution of present-day odonates is wide-ranging, from the Arctic to the tropics, including desert regions.

The adults are easily recognizable. They have two pairs of large, powerful wings that are elongate, membranous, and many-veined. The wings move independently of each other, enabling the insect to fly backward as well as forward. At rest, dragonflies hold their wings outstretched and horizontal to the body, while damselflies usually hold their wings together vertically over the body. The body of odonates is elongate and cylindrical and variously colored, sometimes with brightly colored markings. The head is large with large compound eyes and very small, bristlelike antennae. The mouthparts are adapted for biting and chewing; the scientific name for the order refers to the jaws, which are large and sharply toothed. The legs are long and slender and modified to form a "scoop" with which to catch and hold prey for feeding during flight. The males have appendages at the tip of the abdomen that are used for grasping females when mating.

Although metamorphosis is incomplete, the naiads (or nymphs) are aquatic and very different in appearance from the adults. They are wingless, have either external or internal gills, and use their legs for locomotion rather than for the capture and handling of prey. They are also less conspicuous than the adults: naiads are generally drab in color to match their surroundings. Like the adults, they have a large head with well-developed chewing mouthparts, but their eyes are smaller and the antennae are usually more prominent than those of the adults. Their most distinctive characteristic is a long, double-hinged lower jaw that is quickly thrust forward to capture prey. At maturity, the naiads crawl out of the water onto various objects, where the adults then emerge from the naiad skin, dry their wings, and fly away.

Many species of odonates, in particular, the dragonflies, spend a large portion of their adult life on the wing, surpassing all other insects with their flying skills. They rely on flight to disperse to new areas, to hunt and capture prey, and to search for mates. Most dragonflies mate in flight, although damselflies and some dragonflies mate while perched on objects like the stems of plants. Many species lay their eggs in flight, either by dropping the eggs from above into water or by dipping their abdomen in the water.

The adults are usually found near ponds and streams, although they may travel away from water, some for long distances, either to disperse or to search for prey. Airborne dragonflies are often aggressive in pursuing prey and in challenging intruders, including humans. They usually patrol established routes along the shoreline, alighting on various objects. In contrast, damselflies usually flit about vegetation, their flight slower and more erratic than that of dragonflies. Both adults and naiads are active during the day. Most of the prey taken by adults are flying insects, including other odonates. The naiads live in freshwater and feed on a wide range of aquatic invertebrates, tadpoles, and even small fish.

The presence of odonates is a visible indicator of the diversity and health of aquatic ecosystems. Since both the adults and naiads are predaceous, they play an important role as regulators of insect populations. In addition, because they are so abundant, they serve as an important food source for other wildlife. And, like butterflies, dragonflies have figured in literature, art, and mythology.

Key to Odonata Families

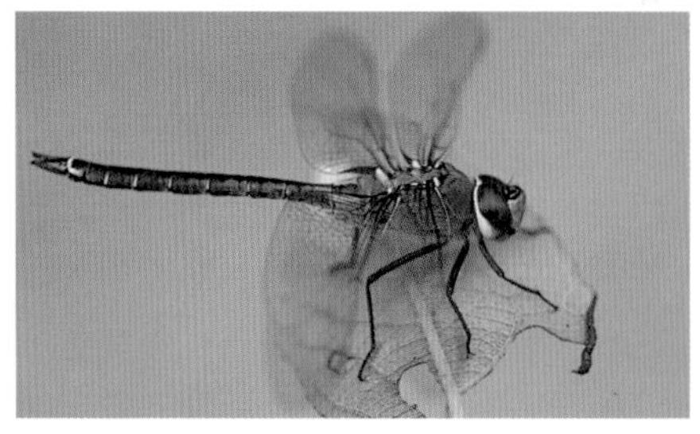

DARNERS
Family Aeshnidae, page 247

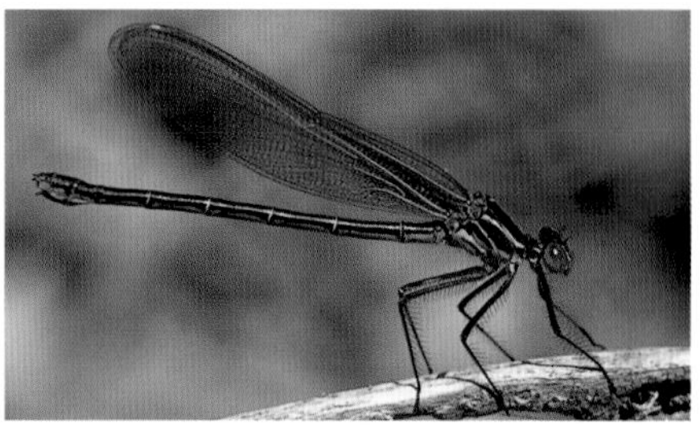

BROAD-WINGED DAMSELFLIES
Family Calopterygidae, page 248

NARROW-WINGED DAMSELFLIES
Family Coenagrionidae, page 249

CLUBTAILS
Family Gomphidae, page 251

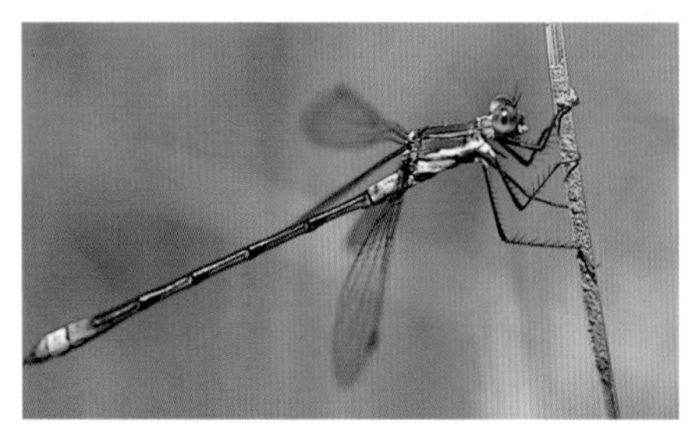

SPREAD-WINGED DAMSELFLIES
Family Lestidae, page 252

COMMON SKIMMERS
Family Libellulidae, page 253

DARNERS Family Aeshnidae

This family includes the largest and most powerful dragonflies in North America. They have a very stout thorax and a long, slender abdomen, which supposedly resembles a darning needle (hence the common name). The body is usually dark brown with green or blue markings, the wings are large and transparent, and the large eyes meet broadly on top of the head. The females have a well-developed ovipositor.

Aeshna multicolor

BLUE-EYED DARNER

ADULT Male, thorax brown with pair of bright blue lines on each side; abdomen coppery brown with bright blue spots and black transverse lines; head and eyes bright blue. Female, may be similar to male; or, thorax coppery brown with pair of light yellow or light green lines on each side, abdomen coppery brown with light yellow or light green spots and black transverse lines. **BODY LENGTH** 72 mm. **FOOD** Adult and naiad: Invertebrates. **FOUND** From British Columbia to California. Adult may be found long distances from water.

Aeshna multicolor, male

Anax junia

COMMON GREEN DARNER

ADULT Male, thorax green; abdomen blue turning to mostly brown posteriorly (in cool weather, abdomen turns purplish brown [as shown]); head with dorsal bull's-eye mark. Female, similar to male but abdomen with green instead of blue, or similar in all aspects. **BODY LENGTH** 79 mm. **FOOD** Adult and naiad: Invertebrates. **FOUND** Common throughout the U.S.

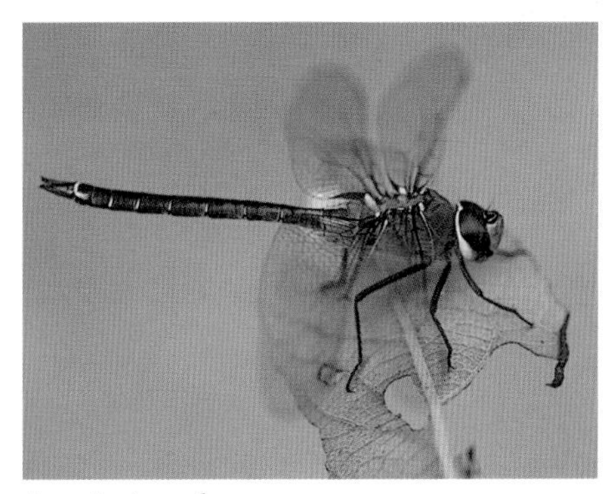

Anax junia, male

BROAD-WINGED DAMSELFLIES Family Calopterygidae

These relatively large damselflies occur along streams. They are the most easily identifiable of the damselflies. Their wings broaden gradually from the base, rather than being stalked, as in other damselfly families. When at rest, the wings are held together vertically above the slender body, which is usually an iridescent black or green. The naiads are slender and long-legged.

Hetaerina americana, male

Hetaerina americana, female

Hetaerina americana

COMMON RUBYSPOT

ADULT Male, thorax dark brown; abdomen bronzy brown; head dark brown; wings with deep red base. Female, thorax medium brown; abdomen bronzy brown; head brown; wings with reddish brown base. **BODY LENGTH** 44 mm. **FOOD** Adult and naiad: Invertebrates. **FOUND** Throughout the U.S.

Most species of North American damselflies belong to this large family, and they are the most common damselflies in the Pacific Northwest. They occur in a variety of habitats; many inhabit marshes or ponds, and others are more common along streams. They are very similar in appearance to the spread-winged damselflies (Lestidae). The body of these small- to medium-sized, delicate damselflies is usually brightly colored; most are black with blue or green markings, and the males are often more brightly colored than the females. The abdomen is long and slender. As the common name of the family suggests, the long, stalked wings are narrow. Most species are feeble flyers, and, at rest, hold the body horizontally with the wings together over the abdomen. The naiads have leaflike gills at the tip of their abdomen.

Argia lugens

SOOTY DANCER

ADULT Male, dark brown (thorax may turn blue with age) with distinct, pale brown ring around each abdominal segment. Female, similar to male except thorax brown with black lines and does not turn blue with age. **BODY LENGTH** 50 mm. **FOOD** Adult and naiad: Invertebrates. **FOUND** Throughout the region.

This is one of the largest damselflies in the region.

Argia lugens, female

Ischnura cervula

PACIFIC FORKTAIL

ADULT Male, thorax black with four blue dorsal spots and blue on sides; abdomen black with blue near tip. Female, similar to male; thorax and head may be pale pink. **BODY LENGTH** 29 mm. **FOOD** Adult and naiad: Invertebrates. **FOUND** Very common throughout the region.

This is one of the easiest damselflies to identify because of the four blue dorsal spots on the thorax. It is normally the first damselfly to emerge in the spring.

Ischnura cervula, male

Ischnura erratica

Zoniagrion exclamationis, mating pair

Ischnura erratica

SWIFT FORKTAIL

ADULT Thorax black with blue bands; abdomen dark brown with dorsal blue patch near tip. **BODY LENGTH** 34 mm. **FOOD** Adult and naiad: Invertebrates. **FOUND** Throughout the region.

The adult flies in the spring.

Zoniagrion exclamationis

EXCLAMATION DAMSELFLY

ADULT Thorax blue and black with dorsal pair of blue exclamation-point marks (not always present in females) in center; abdomen black with blue patch at tip. **BODY LENGTH** 35 mm. **FOOD** Adult: Insects. Naiad: Aquatic invertebrates. **FOUND** Oregon and California.

This is one of the easiest damselflies to identify because of the exclamation points on the thorax.

CLUBTAILS Family Gomphidae

The members of this fairly large family are typically associated with streams or lakes. The adults are large and have a dark-colored body, usually marked with yellow or green. The eyes are widely separated dorsally. The posterior end of the abdomen is often enlarged, most evidently in the males (hence the common name). The females lack an ovipositor. The adults are usually found resting, well camouflaged, on vegetation. The naiads have a wedge-shaped head and thick antennae.

Cordulegaster dorsalis

PACIFIC SPIKETAIL

ADULT Thorax dark brown to black with pair of pale yellow bands on each side; abdomen dark brown to black with pale yellow spot on each segment (other western spiketails have banded abdomens). **BODY LENGTH** 85 mm. **FOOD** Adult and naiad: Invertebrates. **FOUND** Throughout the region.

Cordulegaster dorsalis

Octogomphus specularis

GRAPPLETAIL

ADULT Black; thorax with yellow patches; abdomen with thin, yellow dorsal line; face yellow. **BODY LENGTH** 53 mm. **FOOD** Adult and naiad: Invertebrates. **FOUND** Throughout the region.

Octogomphus specularis

Ophiogomphus bison

BISON SNAKETAIL

ADULT Thorax green with dark brown to black band on each side; abdomen black with yellow dorsal patch and white ventral patch on each segment. **BODY LENGTH** 50 mm. **FOOD** Adult and naiad: Invertebrates. **FOUND** Oregon and California.

Ophiogomphus bison

SPREAD-WINGED DAMSELFLIES Family Lestidae

These damselflies occur mainly around marshes and ponds but can also be found along streams. They are medium to large in size and slender with a long, thin abdomen. As in the narrow-winged damselflies (Coenagrionidae), the long, narrow wings taper to a stalk at the base. These insects are called spread-winged damselflies because, unlike other damselfly families, they hold their body vertically to the substrate and their wings partly outspread when at rest. The naiads have long gills with rounded ends, unlike the narrow-winged damselfly naiads (Coenagrionidae), which have leaflike gills.

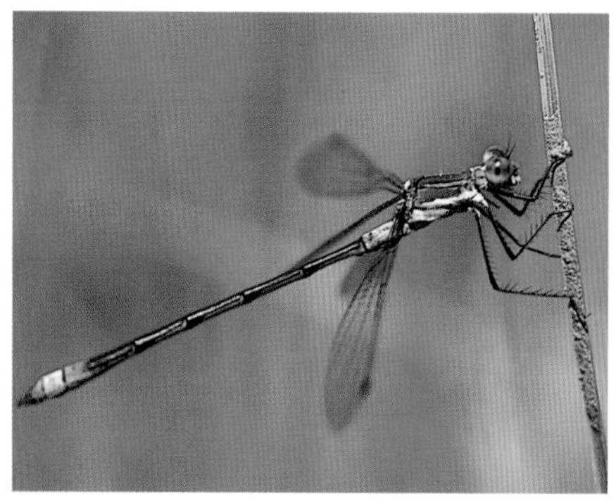

Lestes disjuncta

Lestes dryas

Lestes disjuncta

COMMON SPREADWING

ADULT Thorax blue (turns pruinose blue with age); abdomen iridescent green-black with first two and last two segments pruinose blue. **BODY LENGTH** 40 mm. **FOOD** Adult and naiad: Invertebrates. **FOUND** Throughout the region.

Lestes dryas

EMERALD SPREADWING

ADULT Thorax and abdomen green to blue-green; tip of abdomen often pale blue. **BODY LENGTH** 40 mm. **FOOD** Adult and naiad: Invertebrates. **FOUND** Throughout the region. More common in mountains. Adult found around small ponds with thick vegetation.

COMMON SKIMMERS Family Libellulidae

This is the largest family of Odonata in the world, and many species are quite common. Most species occur at ponds and marshes. The adults in this diverse family are usually medium to large in size, have variously colored bodies, which in many species are brightly colored, and often have distinctly patterned wings. One characteristic shared by all adults is their large eyes in broad contact on top of the head, similar to the darners (Aeshnidae). The short, stout naiads live in shallower, warmer waters than do the naiads of other families and tend to be more active.

Erythemis collocata

WESTERN PONDHAWK

ADULT Male, mostly pruinose blue; face green. Female, green; abdomen with segments edged in black and black dorsal line. **BODY LENGTH** 42 mm. **FOOD** Adult and naiad: Invertebrates. **FOUND** Throughout the region.

Erythemis collocata, female

Libellula lydia

COMMON WHITETAIL

ADULT Male, wings each with broad, dark brown band in middle and dark brown patch at base; abdomen white. Female, wings each with brown patch at base, middle, and tip; abdomen brown. **BODY LENGTH** 46 mm. **FOOD** Adult and naiad: Invertebrates. **FOUND** Throughout the U.S.

Libellula lydia, male

Sympetrum illota, male

Tramea lacerata

Sympetrum illota

CARDINAL MEADOWHAWK

ADULT Male, red; thorax with two faint white spots on each side; wings with red veins. Female, similar to male but not as red. Wings held forward at rest. **BODY LENGTH** 38 mm. **FOOD** Adult and naiad: Invertebrates. **FOUND** Throughout the region.

Tramea lacerata

BLACK SADDLEBAGS

ADULT Black to bluish black, often with yellow spot(s) on dorsal side of abdomen; hindwings each with large black irregular patch at base. **BODY LENGTH** 55 mm. **FOOD** Adult and naiad: Invertebrates. **FOUND** Throughout the region. Although typically near water, adults are strong flyers and may forage far from water.

GRASSHOPPERS, CRICKETS, AND KATYDIDS
Order Orthoptera

Orthoptera (Greek, *ortho,* straight; *ptera,* wings) is a very large, diverse order that contains many large and well-known insects. Recently, the composition of Orthoptera significantly changed: formerly included in this order, cockroaches, mantids, walkingsticks, and earwigs, though sharing many characteristics with grasshoppers, crickets, and katydids, have been placed in other orders. (Note: some entomologists now place cockroaches, along with mantids, in the order Dictyoptera; however, since we include only one species of cockroach in this field guide, it is listed under Orthoptera. The following description of Orthoptera will apply to grasshoppers, crickets, katydids, and cockroaches as a group, unless otherwise specified.)

Orthopterans are distributed throughout the world, but most species are tropical. Approximately 12,500 known species of grasshoppers, crickets, and katydids occur worldwide, with over 1000 species described in the United States and Canada. There are approximately 4000 species of cockroaches, and approximately 50 occur in the United States and Canada, most of them southern in distribution. Cockroaches are among the oldest winged insects, first appearing approximately 350 million years ago.

Orthopterans range in size from small to large but most in the Pacific Northwest are medium-sized. They have bodies that are tough or leathery. Most species are winged with two well-developed pairs of wings, and some are wingless. The scientific name of the order refers to the forewings, which are typically thickened and are not used for flying, but instead function more as covers for the hindwings. The hindwings are membranous and held over the abdomen when at rest. The slender antennae are usually long, sometimes longer than the body. The cerci, if present, are typically short. The mouthparts are of the chewing type.

Orthopterans undergo incomplete metamorphosis. The nymphs are wingless and smaller than the adults but otherwise similar in appearance to the adults.

Grasshoppers, crickets, and katydids are easily recognizable insects. Their robust body is elongate and cylindrical with a prominent, saddle-shaped pronotum. The hind legs are usually well-developed, enabling most grasshoppers and crickets to be outstanding jumpers. When present, the forewings are elongate, and the hindwings are broad and commonly folded fanlike beneath the forewings when at rest. Both sexes usually have auditory organs called tympana located on the abdomen (grasshoppers) or on the front legs (crickets and katydids). The female ovipositor is often greatly elongate.

Grasshoppers, crickets, and katydids are well known for their "singing" ability. The songs of these insects are typically produced by rubbing together one body part against another, usually either the forewings or the hind legs and forewings. Typically, only the males produce sound. The songs play an important role in behavior, mainly for the purpose of attracting the opposite sex. Each species typically has a unique song, so distinctive that grasshoppers and especially crickets and katydids can be identified solely on the basis of their song.

Although grasshoppers, crickets, and katydids are found in many types of habitats, they are most typically associated with grasslands.

Most species feed on plants, although some are predators or scavengers.

Because they are relatively large in size, have a prodigious appetite, and have the ability to increase to great numbers, many species of grasshoppers and crickets can have a profound effect on agricultural fields as well as native vegetation. The destructive nature of huge swarms of grasshoppers are legendary. In the Old Testament of the Bible, grasshoppers are the "locusts" that destroy the crops of Egypt. In the United States, damaging multitudes of grasshoppers are usually limited to the eastern part of the Pacific Northwest and the Great Basin.

Grasshoppers have also been depicted as "bad" insects in other ways. According to folklore, the "lazy" grasshopper who does not prepare for, and consequently does not survive, the winter season is compared to the "hard-working" ant who does prepare and survive. This tale was updated in the movie *A Bug's Life* (Disney/Pixar 1998), where grasshoppers are depicted not only as lazy but also as freeloading "motorcycle gang" members that bully other insects. Crickets, on the other hand, are so well thought of that they have occupied an important place in the traditions of some cultures.

SHORT-HORNED GRASSHOPPERS Family Acrididae

Most grasshoppers in the United States and Canada belong to this large and diverse family. These medium- to large-sized grasshoppers, also called *locusts*, occupy almost all habitats. They differ from most of the other jumping orthopterans by their short, horn-shaped antennae, which are usually less than half the length of the body (hence the common name). Their auditory organs are located on the sides of the abdomen. The hind legs are modified for jumping. Most species are winged and strong flyers. The males sing during the day—most species produce a low, buzzing sound by rubbing their hind legs against the roughened surfaces of their forewings, and others produce a kind of crackling sound by snapping their hindwings in flight. The females have a short ovipositor with which they deposit their eggs in the ground after mating. These grasshoppers feed on plants and can be very destructive when they build up to high numbers, which usually happens in dry habitats.

Arphia conspersa

ORANGE-WINGED GRASSHOPPER

ADULT Above, forewings brown; hindwings commonly orange, but also yellow or red, with black outer margin. **BODY LENGTH** 30 mm. **FOOD** Adult and nymph: Grasses and forbs. **FOUND** Throughout the West.

This species appears in spring to early summer, whereas similar species appear in late summer to fall.

Dissosteira carolinus

MOURNING CLOAK GRASSHOPPER

ADULT Above, forewings brown; hindwings black to dark brown with light yellow outer margin. **BODY LENGTH** 50 mm. **FOOD** Adult and nymph: Plants. **FOUND** Throughout the West. Adults common in areas with sparse vegetation.

The adult hovers in flight, and the male may produce a loud, crackling sound.

Arphia conspersa, posed

Dissosteira carolinus, posed

Schistocerca shoshone

Schistocerca shoshone

GREEN VALLEY GRASSHOPPER

ADULT Green; head and pronotum with yellow dorsal line; hind tibiae pink. **BODY LENGTH** 65 mm. **FOOD** Adult and nymph: Grasses and many other plants. **FOUND** Oregon and California. In late summer, adults congregate on vegetation that is still green and succulent (e.g., willows and plants near seeps).

BROWN-HOODED COCKROACHES Family Cryptocercidae

There is only one species in this family.

Cryptocercus punctulatus

Cryptocercus punctulatus

SCUDDER BROWN-HOODED COCKROACH

ADULT AND NYMPH Reddish dark brown; without wings or wingcovers. **BODY LENGTH** 29 mm. **FOOD** Adult and nymph: Wood. **FOUND** Throughout much of the U.S. Adult and nymph found under bark of dead trees and inside rotting wood (Pete has found groups of adults living together inside large, rotting logs).

This species was previously thought to be the missing link between cockroaches and termites. This linkage is now in dispute, although it is thought these cockroaches may be distantly related to termites. They have a life span of six to seven years.

JERUSALEM OR SAND CRICKETS Family Stenopelmatidae

This very small family commonly occurs west of the Rocky Mountains, mainly along the Pacific Coast. The adults are large, usually brown in color, wingless, with a bulbous head, large abdomen, and distinctly spiny legs. These nocturnal crickets can usually be found under objects during the day. They feed on plants. Although common, this family is poorly known. Accurate identification to species for this family is not yet possible, according to Dr. David Weissman (personal communication), who has been revising the taxonomy of Jerusalem crickets.

Stenopelmatus spp.

ADULT AND NYMPH Shiny brown; abdomen usually banded with black. **BODY LENGTH** 40 mm. **FOOD** Adult and nymph: Plants and invertebrates. **FOUND** Throughout the region but more common in the southern part. Normally active only after dark: can be found moving around on the ground at night.

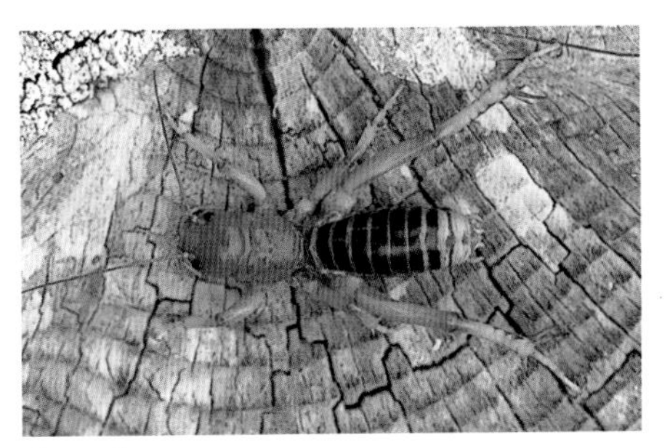

Stenopelmatus spp.

KATYDIDS OR LONG-HORNED GRASSHOPPERS

Family Tettigoniidae

These large, nocturnal insects can usually be recognized by their very long threadlike antennae, which are usually much longer than the body. Adults are usually winged and have forewings broader than those of other grasshoppers. The hind legs are modified for jumping. Body coloration ranges from dark brown to green, with most being green. The auditory organs are located at the base of the front tibiae. Most species are highly vocal, and many are known for their songs: males produce stridulant sounds by rubbing together the roughened surfaces of their forewings. The females have a long, flat, swordlike ovipositor, with which, in many species, they insert their eggs into plant tissues. Most species feed on plants, but a few prey on other insects.

Anabrus simplex

MORMON CRICKET

ADULT AND NYMPH Dark brown to black; pronotum elongate, usually hiding reduced wings. **BODY LENGTH** 60 mm. **FOOD** Adult and nymph: Plants and invertebrates. **FOUND** Eastern part of the region.

In the summer these crickets are an impressive sight as great numbers of them move across roads, which can become slick with their smashed bodies (the live crickets turn into cannibals, eating their dead, as well as living, brethren).

Anabrus simplex

STONEFLIES Order Plecoptera

Plecoptera (Greek, *pleco*, plaited; *ptera*, wings) is a small order of aquatic insects, with approximately 1550 known species worldwide. They are found near streams and lakes throughout the world, mostly occurring in the cooler regions. Approximately 460 species have been described in the United States and Canada; they occur in all the regions but most abundantly in the mountainous areas of the northern section.

The adults, which are small to large in size, have two pairs of long, many-veined, membranous wings, with the hindwings slightly shorter and much broader than the forewings. The name of the order refers to the way the resting adult holds its wings flat, with the hindwings folded fanlike under the forewings, almost completely covering the abdomen. Although the wings are well developed, stoneflies are such reluctant flyers that they usually run when threatened. The body of the adult is flattened, elongate, leathery, and usually brown or black. Threadlike cerci are present and may be short or long. The antennae are long, slender, and multisegmented. Although the mouthparts are of the chewing type, they are often reduced and nonfunctional. The adults are close to orthopterans in wing venation and other morphological characteristics but are distinguishable from them and the mayflies by their long antennae and wing shape; they are different from other winged insects by their widespread legs and caudal appendages.

Stoneflies undergo incomplete metamorphosis. The naiads (or nymphs) are aquatic. They have broad, elongate, flattened bodies, well-developed legs set wide apart, antennae shorter than those of the adult, and two long cerci. They require well-aerated freshwater from which they obtain oxygen using hairlike or branched gills that are often arranged in tufts on the thorax and near the base of the legs. The mouthparts are of the chewing type. The naiads are very similar to the adults in appearance; in fact, adults retain much of the shape of the naiads, usually just adding wings and discarding the external gills. Stonefly naiads closely resemble mayfly naiads (mayflies are not included in this field guide) but can be differentiated from them by the fact that mayfly naiads usually have three cerci, leaflike gills that are located along the sides of the abdomen, and shorter antennae.

Stoneflies are found in or near usually fast-flowing streams or on rocky lakeshores. The adults typically remain near the water from which they emerged, commonly resting on shoreline vegetation and rocks. The naiads can be found crawling along the bottom of a stream or along the shore of a lake, but more often are found under objects such as stones (hence the common name of the order). The adults of many species do not feed and are short-lived; the few that do eat are plant feeders. The naiads of many species feed on detritus, while others feed either on plants or a wide variety of both plant and animal matter; those of a few families are predaceous.

Stoneflies are an important source of food for fish and other vertebrates. They make excellent trout bait for anglers, who use them as live bait or use artificial baits that mimic stoneflies. Stonefly naiads are considered indicators of water quality because they can survive only in clean, highly oxygenated water; therefore polluted waters significantly threaten stonefly populations.

GIANT STONEFLIES Family Pteronarcyidae

This small family contains the largest of the stoneflies. The adults have a brown or gray body, black-veined wings, and long antennae. They are nocturnal and often attracted to lights in late spring and early summer when they emerge. They do not feed. The naiads bear gills on the first three abdominal segments. They inhabit streams and feed on plant material.

Pteronarcys californicus

CALIFORNIA PTERONARCYS

ADULT Pronotum orange or red with two large, black spots. **BODY LENGTH** 45 mm. **FOOD** Naiad: Aquatic plants. **FOUND** Throughout the region. Naiad, in clear, fast-moving streams.

The adult generally flies in spring and early summer.

Pteronarcys californicus

PART TWO
NON-INSECT INVERTEBRATE ORDERS

Key to Non-insect Invertebrate Orders

MITES & TICKS
Order Acari, page 265

SPIDERS
Order Araneae, page 268

POLYDESMID MILLIPEDES
Order Polydesmida, page 274

LAND SNAILS & SLUGS
Order Stylommatophora, page 275

ARACHNIDS Class Arachnida

Arachnids (Greek, *arachni*, spider), which include spiders, ticks, mites, and scorpions, are the largest non-insect class of arthropods. More than 75,000 species are known to occur worldwide, with 4000 species described in North America. This class is older than that of insects, with the first arachnids appearing over 400 million years ago.

Arachnids, most of which are terrestrial, differ from insects in several ways. The body is divided into two distinct regions, the cephalothorax (combined head and thorax) and the abdomen. In most arachnids, the two regions are distinct. They have four (or less) pairs of walking legs, and simple, rather than compound, eyes, and lack antennae. They also have a pair of jawlike or fang-bearing appendages, called *chelicerae*, in front of the mouth, and a pair of leglike pedipalps between the chelicerae and the first pair of walking legs. While many insects are plant feeders, most arachnids are predaceous, feeding on invertebrates, primarily insects.

MITES AND TICKS Order Acari (Acarina)

Acari (Greek, *acar*, mite, tiny) is a very large order of minute to small arachnids, with more than 30,000 known species of mites and approximately 1000 known species of ticks worldwide. It is thought that these animals may actually surpass insects in total number of species, but because they are so small, a great many are still to be discovered. In fact, many that are known are almost impossible to identify in the field below the suborder level.

Mites and ticks have a body that is usually oval and compact, with one to four pairs of legs. Unlike most arachnids, though, there is generally little or no differentiation of the two body regions. The legs are attached to the head region. The abdomen contains the respiratory, digestive, and reproductive systems. Ticks are typically larger than mites.

The newly hatched young, called *larvae*, have only three pairs of legs. They acquire the fourth pair after the first molt. The instars between the larval and adult stages are called *nymphs* and are similar to the adults.

This group includes both terrestrial and aquatic forms. Mites are abundant in soil and organic debris, where they usually outnumber other arthropods. Many species of mites are either predaceous, plant feeders, or scavengers. Some form galls. Some are parasites, particularly in the larval stages, of both vertebrates and invertebrates; most of the parasitic forms are external parasites.

Mites in general play an important part in regulating insect populations above and below ground, and in promoting soil fertility by breaking down organic material. Predaceous mites are used commercially as biological control agents. Some species are considered to be significant pests in agriculture and nursery plant production.

All ticks in both the adult and immature stages are external blood-feeding parasites of invertebrates and vertebrates (mainly mammals, birds, and reptiles). Some species transmit vertebrate (including human) diseases.

ERIOPHYID MITES Family Eriophyidae

This large family includes species that are commonly called *gall, rust, bud,* or *blister mites.* These mites are so small that they are almost invisible to the naked eye; their features usually cannot be seen without a hand lens. They are elongate, soft-bodied and wormlike, and unique among mites in having only two pairs of legs. Most species feed on leaves; a few species form small pouch-like galls on leaves, while others attack buds. Many species are major pests of orchard trees and other cultivated plants.

Aculops tetanothrix, galls

Phytophus emarginata, galls

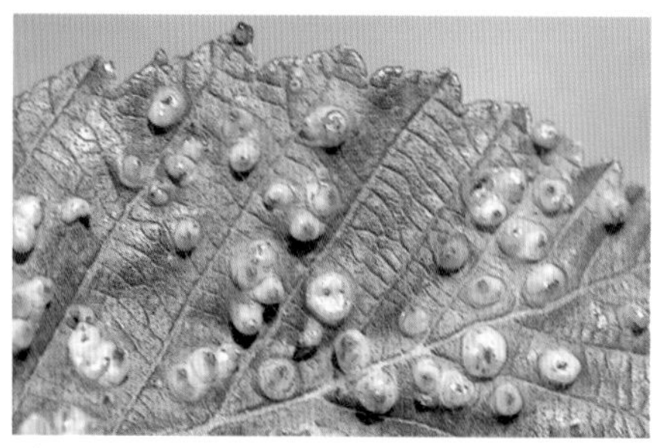

Phytophus laevis, galls

Aculops tetanothrix

WILLOW GALL

GALL Bead-shaped and irregularly round; red when exposed to sun, otherwise yellow; on surface of leaf. Yellow mite inside. **GALL WIDTH** 3 mm. **FOOD** Adult and immature: Coastal willow (*Salix hookeriana*) and other willows (*Salix* spp.). **FOUND** Common throughout the natural range of the host.

High numbers of galls can distort the leaf.

Phytophus emarginata

CHERRY POUCH GALL

GALL Red or yellow pouch; on upper surface of leaf. **GALL WIDTH** 2 mm. **FOOD** Adult and immature: Bitter cherry (*Prunus emarginata*) and other *Prunus* spp. **FOUND** Throughout the region.

Phytophus laevis

ALDER BEAD GALL

GALL Beadlike; pale green to red; on upper surface of leaf. **GALL WIDTH** 2 mm. **FOOD** Adult and immature: Red alder (*Alnus rubra*), white alder (*A. rhombifolia*), and other alders (*Alnus* spp.). **FOUND** Throughout the region.

HARD TICKS Family Ixodidae

The members of this family have a hard dorsal plate, called a *scutum* (from which they get their common name), and mouthparts that protrude anteriorly and are visible from above—these two taxonomic features differentiate these ticks from the soft ticks (which are not included in this field guide). They usually parasitize two or three hosts during their development. The males of some species do not feed. Hard ticks are significant vectors of diseases such as tularemia and Lyme disease.

Dermacentor variabilis

AMERICAN DOG TICK

ADULT Dark reddish brown with white dorsal markings. **BODY LENGTH** 5 mm. **FOOD** Adult and immature: Mammals, including humans (parasitic). **FOUND** Throughout the U.S.

The tick in the photograph is on Pete's pant leg looking for a meal!

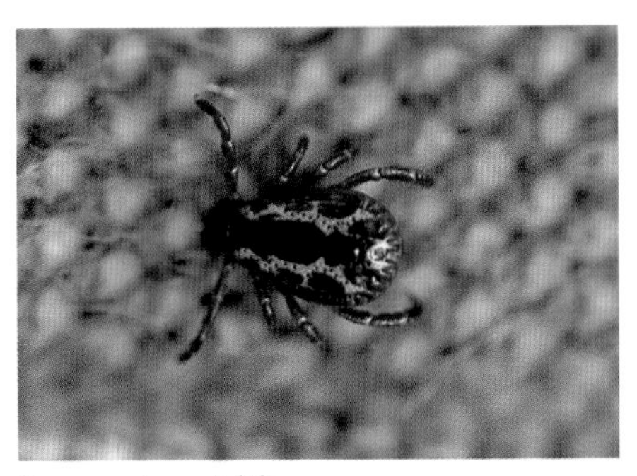

Dermacentor variabilis

SPIDERS Order Araneae

Araneae (Latin, *arane*, spider) is a very large order comprising more than 35,000 described species worldwide, including approximately 3000 species in North America. It is believed that many more species are undocumented. Spiders are widely distributed throughout the world and occur in almost every kind of habitat, from underground caves to the highest mountain ranges (even on Mount Everest). They are often very abundant; in one study, researchers found approximately 2,200,000 spiders in a one-acre plot of grassland. Spiders exhibit the largest variety of shape, color, and behavior of all the arachnids.

Spiders are easily identifiable by the presence of four pairs of legs and the two body regions that, unlike those of other arachnids, are not broadly joined together but are connected by a narrow waist. Spiders also differ from all other arachnids in having an unsegmented, or a more or less saclike, abdomen. The cephalothorax is covered dorsally by a shieldlike covering, called a *carapace*. Most species have eight eyes, but some have fewer, and a few have none. The number and position of the eyes are often important features in identifying to family. The pair of small jaws ends in fangs. All spiders have silk glands, called *spinnerets*, located at the tip of the abdomen below the anus.

With few exceptions, all spiders have poison glands. The venom, expelled through a duct in the fangs, is used to paralyze and kill prey or as a means of defense. Only a few species produce venom powerful enough to be a danger to humans. Furthermore, most species are shy and normally will not bite if handled carefully.

Most spiders lay their eggs in silken sacs. Some place their egg sac in a web, while others attach it to plants or other objects, and still others carry the sac around until the eggs hatch. Spiders undergo very little metamorphosis during their development; the young, called *spiderlings*, resemble the adults. In many species, the two sexes differ greatly in size, with the female being larger than the male.

The silk that spiders produce from their spinnerets is of various types and serves different functions. The most important of these is the *dragline*, a single strand of silk many spiders play out behind them as they move around. It is fastened to the substrate at intervals by means of attachment disks. Spiders use the dragline as a bridge between two objects and to escape danger. Silk is also used to spin webs (or snares) or line burrows or other kinds of retreats, to form the females' egg sacs and the males' sperm webs, and to wrap prey into cocoons to be fed on later. The spiderlings, and sometimes adult males, of many species commonly engage in an aerial dispersal method called *ballooning*. The spider climbs to a high point, for example, on top of a rock or the tip of a blade of grass, and faces into the wind. It then raises its abdomen and releases several long silken threads which are carried into the air. When enough silk has been played out, the spider becomes airborne, its silken parachute carrying the spider away to other areas.

Most spiders lead solitary lives, but a few species are social, sharing their webs and prey. Not all spiders spin webs: some live in silk-lined burrows, whereas others have no retreat at all. The webs greatly vary in structure and are characteristic of the species. Most species are active during the day, using their keen eyesight to hunt for their prey, while other species are either active at night or live under objects and rarely venture forth.

All spiders are predaceous and feed on invertebrates, primarily insects; a few large species may prey on small vertebrates. Spiders, especially the young, may feed on other spiders. Typically, a spider first paralyzes the prey with its poisonous bite, after which it injects digestive enzymes, which break down the tissues of the immobilized prey. The spider then sucks up the liquified tissues, leaving only an empty shell. Some species also scavenge.

Because of their great numbers and predatory habits, spiders play a significant role in ecological processes, although this role is noticeably underappreciated by the general public. They are an important component in the regulation of invertebrate populations, and, in turn, an important food source for other wildlife, particularly birds and wasps.

ORB WEAVERS Family Araneidae

This is a very large family of common spiders that have a wide distribution. They greatly vary in size, shape, and color and are often distinctly colored and patterned. The males, less likely to be encountered, are typically much smaller than the females. All are distinguishable by their eight eyes that are arranged in two horizontal rows of four eyes each.

These spiders can also be easily recognized just by their large webs, which almost all species construct in the form of a spiraling orb. In spinning the web, the spider first lays down support lines, which radiate outward from the center and give the web strength. The spider then spins in the middle of the web the sticky spiral threads, which are used to snare and hold prey. Some species construct a retreat at a distance from their orb web, while others remain for the most part at the center of their orb, usually hanging head down, patiently awaiting their next victim. The male spider commonly spins its own orb web in an outlying part of the female's web. Many species replace the entire web daily, spinning the new web usually in the early evening.

Araneus diadematus

CROSS SPIDER

ADULT Female, light brown to darker reddish brown; abdomen with dark, dorsal markings, and distinct, lateral row and less distinct, medial row of white spots that often form a cross. **BODY LENGTH** 13 mm. **FOOD** Invertebrates. **FOUND** Throughout the region.

This species was introduced from Europe.

Araneus diadematus, female

Araneus trifolium, female

Araneus trifolium

SHAMROCK SPIDER

ADULT Female, abdomen variably colored but often reddish brown with distinctive, white, dorsal spots. **BODY LENGTH** 19 mm. **FOOD** Invertebrates. **FOUND** Throughout the U.S. In late summer look for large webs in tall grass or brushy areas. The female, if not visible on the web itself, will normally be in or under a shelter of leaves and silk that is connected to the web by a strand of silk. Because the male is small and timid, he is not usually visible.

One of the most common large orb weavers in the region.

Argiope aurantia, female

Argiope aurantia, spiderlings

Argiope aurantia

BLACK-AND-YELLOW ARGIOPE

ADULT Female, abdomen boldly colored yellow and black. **BODY LENGTH** 28 mm. **FOOD** Invertebrates. **FOUND** Common throughout the U.S.

This spider is commonly known as the *Halloween spider*, which is appropriate since by November the female's strikingly colored body is swollen with eggs. In the spring the spiderlings, after emerging from the egg case, stay together in clumps for a day or two and then disperse.

Argiope trifasciata, female

Argiope trifasciata

BANDED GARDEN ARGIOPE

ADULT Female, abdomen banded yellow and white. **BODY LENGTH** 25 mm. **FOOD** Invertebrates. **FOUND** Throughout the U.S.

Larinioides cornutus

FURROW SPIDER

ADULT Female, abdomen gray with dark brown dorsal markings (lighter than *Larinioides patagiatus*). **BODY LENGTH** 14 mm. **FOOD** Invertebrates. **FOUND** Throughout the U.S.

Also known as *Nuctenea cornuta*.

Larinioides cornutus, female

Larinioides patagiatus

ADULT Male, light brown. Female, abdomen brown (darker than *Larinioides cornutus*) with posterior half solid dark brown. **BODY LENGTH** 11 mm. **FOOD** Invertebrates. **FOUND** Throughout most of the northern U.S.

Also known as *Nuctenea patagiatus*.

Larinioides patagiatus, male and female

Metepeira grinnelli

ADULT Female, brown; abdomen with dark gray bands. **BODY LENGTH** 7 mm. **FOOD** Invertebrates. **FOUND** Range unknown, but Pete has collected this species in Oregon and California.

This species' web is very distinctive: the center is a string of debris that consists of dead insects and plant material held together with webbing. The female sits at the bottom of this debris string.

Metepeira grinnelli, female on debris string

JUMPING SPIDERS Family Salticidae

This is a very large family of small- to medium-sized spiders. They have a short, stout body with short, stout legs. The entire body is often densely covered with hairs or scales that are usually brightly colored or iridescent. These spiders get their common name from their ability to leap great distances; they can also move sideways or backward with equal ease. They have the sharpest vision of all spiders. The eight eyes, which cover the entire length of the head, are arranged in three distinct rows. The spiders make use of their acute vision to hunt for prey in daylight. They are most active during warm, sunny weather. They stalk their prey by creeping toward it until they are a short distance away, and then pounce, grabbing the prey with their forelegs. As they jump onto the victim, the spider's spinnerets emit a dragline that is anchored to the substrate. Jumping spiders do not spin webs, but most construct closely woven, saclike, silken retreats for molting, hibernating, and spending the night on plants or under bark, stones, or plant debris.

Phidippus clarus

Phidippus johnsoni

Phidippus clarus

ADULT Abdomen black with narrow, white anterior fringe, and two broad, red dorsal bands between which are two to three pairs of white spots. **BODY LENGTH** 5 mm. **FOOD** Invertebrates. **FOUND** Throughout the region, commonly on plants, rocks, wood fences, and buildings.

Phidippus johnsoni

JOHNSON'S JUMPING SPIDER

ADULT Black; abdomen with broad, red, U-shaped dorsal band. **BODY LENGTH** 5 mm. **FOOD** Invertebrates. **FOUND** Throughout the region, commonly on plants, rocks, wood fences, and buildings.

Salticus scenicus

ZEBRA JUMPING SPIDER

ADULT Black to reddish brown; abdomen with white stripes. **BODY LENGTH** 5 mm. **FOOD** Invertebrates. **FOUND** Throughout the region, commonly on rocks, plants, and buildings.

Salticus scenicus eating fly

CRAB SPIDERS Family Thomisidae

These spiders are distinctly crablike in appearance and movement, hence the common name of the family. They typically have a short, broad body, which in many species is brightly colored, and legs that are held outstretched to the sides like crabs. The first and second pairs of legs are often much stouter and longer than the third and fourth pairs. These spiders can move forward, sideways, and backward. The eight small, dark eyes are usually located on raised bumps in two backwardly curving rows of four eyes each. Crab spiders do not spin webs, retreats, or overwintering nests but wander about on the ground or on plants in search of prey or lie in ambush, usually on flowers. In some species, the male may cover his potential mate with loose silken webbing and tie her down.

Misumena vatia

FLOWER CRAB SPIDER

ADULT Female, white or yellow; abdomen usually with red spot or streak on each side. **BODY LENGTH** 9 mm. **FOOD** Invertebrates. **FOUND** Very common throughout the U.S. Often in or on a flower.

Misumena vatia, female

MILLIPEDES Class Diplopoda

Diplopoda (Greek, *diplo*, two; *pod*, foot) are elongate, wormlike arthropods with many legs: most millipedes have 30 or more pairs of legs, and most of the body segments have two pairs of legs per segment. The body is cylindrical or flattened, and the antennae are short. The immature are similar in appearance to the adults. Millipedes are usually found in damp places—under leaves, stones, or boards or in moss, rotting wood, or soil. Most are scavengers and feed on decaying plant material. Many species emit an unpleasant-smelling fluid through openings on the sides of the body. This fluid is potent enough to kill any insect placed in a container with a millipede. It has been found that in some species the fluid contains hydrogen cyanide.

POLYDESMID MILLIPEDES Order Polydesmida

Polydesmida (Greek, *poly*, many; *desmi*, band) is the largest order of millipedes. These millipedes have a body that is flattened dorsoventrally and keeled laterally. The body segments have two pairs of legs each, except for the first and last segments, which are legless, and the second and fourth segments, each of which have a single pair. The segments bearing two pairs of legs are a series of hardened rings connected by soft tissue. The eyes are much reduced or absent.

■ POLYDESMOID MILLIPEDES Family Polydesmoidae

The body of millipedes in this family is distinctly flattened dorsoventrally. Most species have twenty body segments and are blind. All species feed on plants, including rotting vegetation.

Harpaphe haydeniana

Harpaphe haydeniana

ADULT Above, black with row of light yellow spots on each side. **BODY LENGTH** 25 mm. **FOOD** Adult and immature: Organic matter. **FOUND** Throughout the region.

May be called *cyanide millipede*. These animals can be so numerous that their bodies can clog outdoor drains after having been washed out of the forest humus by heavy rains. When handled, they release a liquid that stains the skin.

SNAILS AND ALLIES Class Gastropoda

Gastropoda (Greek, *gastro*, stomach, belly; *pod*, foot) is the largest group of mollusks (phylum Mollusca [Latin, *mollusc*, soft, shellfish]), varying in form from snails with complex, twisted shells to shell-less members such as slugs and sea hares. Most gastropods are marine, but others are adapted to live in freshwater or terrestrial habitats; the only group of mollusks that live on land are in this class.

Like other mollusks, gastropods characteristically have a soft body with a fleshy, muscular foot and a fleshy body covering called the *mantle*. The foot functions in locomotion and/or capturing prey. Gastropods have a large, extendable foot on which they glide along, and which lies below most of the digestive organs, thus giving rise to the scientific name of the class. The mantle functions in respiration, sensory reception, and waste disposal. In some mollusks, the mantle produces the external shell. The space enclosed by the mantle is called the *mantle cavity*, and it houses either gills or a respiratory membrane, or lung. Marine gastropods typically have well-formed feathery gills, while terrestrial and freshwater snails and slugs have a lung. Another feature that gastropods share with other mollusks is the *radula*, a rasping, tonguelike structure, which is used to scrape up food.

LAND SNAILS AND SLUGS Order Stylommatophora

The members of this order are the only mollusks that live on land. They are among the most successful and diverse animal groups in terrestrial ecosystems; however, little is known about their biology. They have long been important to human societies as vectors of parasites, agricultural pests, food, currency in trade, tools, and decorative items. Because of their feeding activity, most land snail and slug species play an important role in nutrient cycling.

Family Bradybaenidae

Little is known about this snail family, which has no common name. The shell is usually banded, but greatly varies in shape, from globose to conical to flattened. Other characteristics used to identify this family are too technical to be included in this field guide.

Monadenia fidelis

Monadenia fidelis

PACIFIC SIDEBAND SNAIL

ADULT Shell reddish brown with yellow band; foot (body) black and covered with coppery pink bumps. **BODY LENGTH** 35 mm. **FOOD** Herbaceous plants. **FOUND** Quite common throughout the region, usually in or near forests.

LANCETOOTH SNAILS Family Haplotrematidae

This snail family comprises only two genera. The characteristics used to identify this family refer to the snail's radula and other structures that are not within the scope of this field guide.

Haplotrema vancouverense

ROBUST LANCETOOTH SNAIL

ADULT Shell pale brown with slight green tint and flattened laterally; foot (body) white. **BODY LENGTH** 21 mm. **FOOD** Snails. **FOUND** Throughout the region.

Since this snail preys on other snails, it is considered a friend to gardeners.

Haplotrema vancouverense

Haplotrema vancouverense eating smaller snail

Glossary

Abdomen—the third main body segment, posterior to the thorax.

Apical—at the end or tip.

Cerci (singular, cercus)—a pair of sensory appendages at the posterior end of the abdomen.

Chrysalis—the pupa of a butterfly.

Cocoon—the silken case in which a larva pupates.

Dorsal—top or uppermost; pertaining to the back or upper side of the body.

Dorsoventral—from top to bottom; from the upper surface to the lower surface.

Femur—the third segment of the leg.

Honeydew—a soluble sugar converted from plant starch.

Inquiline—an animal that lives in the home of another species.

Instar—the developmental stage of an insect that is between molts.

Larva (plural, larvae)—the immature stage between the egg and pupa of insects having complete metamorphosis.

Larviform—shaped like a larva; wormlike.

Lateral—on the side.

Medial—in the middle.

Metamorphosis—the transformation of an immature insect into an adult. Complete metamorphosis involves a pupal stage; in incomplete, or simple, metamorphosis the immature stage(s) essentially resembles the adult stage.

Molt—the process of shedding the outer body covering to permit growth or metamorphic change.

Naiad—the aquatic, gill-breathing immature stage (nymph) of some orders.

Nymph—the immature stage between the egg and adult of insects having incomplete metamorphosis.

Ovipositor—the egg-laying structure at the tip of the female's abdomen.

Parthenogenesis—asexual reproduction in which eggs develop without fertilization.

Pheromone—a chemical emitted by an individual that causes a reaction in other individuals.

Proleg—one of the leglike abdominal appendages of certain insect larvae that is used for walking.

Pronotum—the saddle-shaped or flattened area between the head and base of the wings.

Pruinose—having a pale gray powdery coating.

Pupa (plural, pupae)—the transformational stage between the larva and the adult in insects with complete metamorphosis.

Raptorial—adapted for catching prey.

Scutellum—the triangular area posterior to the pronotum (in Coleoptera, Hemiptera, and Homoptera).

Spiracle—a breathing pore.

Suture—the line of juncture of the wingcovers.

Tarsus (plural, tarsi)—the leg segment beyond the tibia.

Thorax—the second main body segment, between the head and abdomen.

Tibia (plural, tibiae)—the fourth segment of the leg, between the femur and the tarsus.

True leg—one of the larval thoracic legs.

Tubercle—a small, rounded protuberance.

Urticating—stinging.

Vein—a thickened line in the wing.

Ventral—lower or underneath; pertaining to the underside of the body.

Wingcover (=elytron; plural, elytra)—the thickened, leathery, or hard front (fore) wing.

Bibliography

Acorn, J. 2001. *Bugs of Washington and Oregon.* Edmonton, Canada: Lone Pine Publishing.

Anderson, R. S., and S. B. Peck. 1985. The carrion beetles of Canada and Alaska: Coleoptera: Silphidae and Agyrtidae. The Insects and Arachnids of Canada, Part 13. Publication 1778, Biosystematics Research Institute, Research Branch, Agriculture Canada, Ottawa, Ontario.

Arnett, R. H., Jr. 1993. *American Insects: A Handbook of the Insects of America North of Mexico.* Gainesville, Florida: The Sandhill Crane Press.

Arnett, R. H., Jr., and R. L. Jacques, Jr. 1981. *Simon and Schuster's Guide to Insects.* New York: Simon and Schuster.

Barker, G. M., ed. 2001. *The Biology of Terrestrial Molluscs.* Wallingford, England: CABI Publishing.

Biggs, K. 2000. *Common Dragonflies of California: A Beginner's Pocket Guide.* Sebastopol, California: Azalea Creek Publishing.

Borror, D. J. 1960. *Dictionary of Word Roots and Combining Forms.* Mountain View, California: Mayfield Publishing.

Borror, D. J., and D. M. DeLong. 1971. *An Introduction to the Study of Insects.* New York: Holt, Rinehart and Winston.

Covell, C. V., Jr. 1984. *A Field Guide to the Moths of Eastern North America.* Boston: Houghton Mifflin.

Dean, M. B. 1979. *The Natural History of Pterotus Obscuripennis Leconte (Lampyridae, Coleoptera).* Master's thesis, Humboldt State University, Arcata, California.

Dillon, E. S., and L. S. Dillon. 1972. *A Manual of Common Beetles of Eastern North America.* 2 vols. New York: Dover Publications.

Dunkle, S. W. 2000. *Dragonflies Through Binoculars: A Field Guide to Dragonflies of North America.* Oxford: Oxford University Press.

Eichlin, T. D. 1975. A guide to the adult and larval Plusiinae of California (Lepidoptera: Noctuidae). Occasional Papers in Entomology, No. 21, Division of Plant Industry Laboratory Services, California Department of Food and Agriculture, Sacramento, California.

Essig, E. O. 1958. *Insects and Mites of Western North America.* New York: Macmillan.

Fender, K. M. 1970. Ellychnia of western North America (Coleoptera: Lampyridae). *Northwest Science* 44: 31–43.

Fitzgerald, T. D. 1995. *The Tent Caterpillars.* Ithaca, New York: Comstock Publishing.

Foote, R. H., F. L. Blanc, and A. L. Norrbom. 1993. *Handbook of the Flies (Diptera: Tephritidae) of America North of Mexico.* Ithaca, New York: Comstock Publishing.

Furniss, R. L., and V. M. Carolin. 1977. Western forest insects. Miscellaneous Publication 1339, U.S. Forest Service.

Glassberg, J. 2001. *Butterflies Through Binoculars: the West: A Field Guide to the Butterflies of Western North America.* New York: Oxford University Press.

Gordon, R. D. 1985. The Coccinellidae (Coleoptera) of America north of Mexico. *Journal of the New York Entomological Society* 93:1–912.

Hamilton, K. G. A. 1982. The spittlebugs of Canada. Homoptera: Cercopidae. The Insects and Arachnids of Canada, Part 10. Publication 1740, Biosystematics Research Institute, Research Branch, Agriculture Canada, Ottawa, Ontario.

Hatch, M. 1957–1971. *The Beetles of the Pacific North West.* 6 vols. Seattle: University of Washington Press.

Headstrom, R. 1973. *Spiders of the United States.* New York: A. S. Barnes and Company.

Hickman, J. C., ed. 1993. *The Jepson Manual: Higher Plants of California.* Berkeley: University of California Press.

Hogue, C. L. 1993. *Insects of the Los Angeles Basin*. Los Angeles: Natural History Museum of Los Angeles County.

Holland, W. J. 1922. *The Moth Book: A Popular Guide to a Knowledge of the Moths of North America*. Garden City, New York: Doubleday. Reprint, New York: Dover, 1968.

Hooper, J. 2002. *Of Moths and Men*. New York: Norton.

Kaston, B. J. 1953. *How To Know the Spiders*. Dubuque, Iowa: W. C. Brown.

Keifer, H. H., E. W. Baker, T. Kono, M. Delfinado, and W. E. Styer. 1982. An illustrated guide to plant abnormalities caused by eriophyid mites in North America. Agriculture Handbook No. 573, Agricultural Research Service, U.S. Department of Agriculture.

Kettlewell, B. 1973. *The Evolution of Melanism*. Oxford: Clarendon Press.

Larew, H., and J. Capizzi. 1983. *Common Insect and Mite Galls of the Pacific Northwest*. Studies in Entomology Number Five. Corvallis, Oregon: Oregon State University Press.

Linsley, E. G., and J. A. Chemsak. 1997. *The Cerambycidae of North America, Part 8: Bibliography, Index, and Host Plant Index*. Vol. 117, California University Publications in Entomology. Berkeley: University of California Press.

Manolis, T. 2003. *Dragonflies and Damselflies of California*. California Natural History Guides No. 72. Berkeley: University of California Press.

Miller, J. C. 1995. Caterpillars of Pacific Northwest forests and woodlands. Publication FHM-NC-06-95, U.S. Forest Service National Center of Forest Health Management, Morgantown, West Virginia.

Miller, J. C., and P. C. Hammond. 2000. Macromoths of Northwest forests and woodlands. Publication FHTET-98-18, Forest Health Technology Enterprise Team: U.S. Forest Service, Morgantown, West Virginia; U.S. Geological Survey, Corvallis, Oregon; and Cooperative Forest Ecosystem Research, Corvallis, Oregon.

Milne, L., and M. Milne. 1980. *The Audubon Society Field Guide to North American Insects and Spiders*. New York: Alfred A. Knopf.

Needham, J. G., M. J. Westfall, Jr., and M. L. May. 2000. *Dragonflies of North America*. Gainesville, Florida: Scientific Publishers.

Osborne, K. H. 1995. Biology of *Proserpinus clarkiae* (Sphingidae). *Journal of the Lepidopterists' Society* 49(1): 72–79.

Paulson, D. 1999. *Dragonflies of Washington*. Seattle: Seattle Audubon Society.

Poole, R. W., and P. Gentili, eds. 1996–1997. *Nomina Insecta Nearctica: A Check List of the Insects of North America*. 4 vols. Rockville, Maryland: Entomological Information Services.

Powell, J. A., and C. L. Hogue. 1979. *California Insects*. Berkeley: University of California Press.

Pyle, R. M. 1981. *The Audubon Society Field Guide to North American Butterflies*. New York: Alfred A. Knopf.

——. 2002. *The Butterflies of Cascadia*. Seattle: Seattle Audubon Society.

Russo, R. A. 1979. *Plant Galls of the California Region*. Pacific Grove, California: Boxwood Press.

Scott, J. A. 1986. *The Butterflies of North America: A Natural History and Field Guide*. Stanford, California: Stanford University Press.

Slater, J. A., and R. M. Baranowski. 1978. *How to Know the True Bugs (Hemiptera—Heteroptera)*. Dubuque, Iowa: W. C. Brown.

Stehr, F. 1987. *Immature Insects*. 2 vols. Dubuque, Iowa: Kendall/Hunt.

Stewart, B. 1997. *Common Butterflies of California*. Point Reyes Station, California: West Coast Lady Press.

Tiemann, D. L. 1967. Observations on the natural history of the western banded glowworm *Zarhipis integripennis* (Le Conte) (Coleoptera: Phengodidae). *Proceedings of the California Academy of Sciences* 35(12): 235–263.

Wallace, R. A., G. P. Sanders, and R. J. Ferl. 1991. *Biology: The Science of Life*. 3d ed. New York: HarperCollins Publishers.

Index

Judy Haggard holds a BA and MA in biology and works as a consulting wildlife biologist.

Peter Haggard holds a BS in wildlife management and has worked as an agricultural inspector in California for more than 30 years. During that time he has collected, photographed, and identified thousands of insects of the Pacific Northwest and maintained a database of hundreds of insect species.